AF662068

Biotechnological Methods of Pollution Control

Biotechnological Methods of Pollution Control

Anand Prakash

RANDOM

RANDOM PUBLICATIONS

NEW DELHI (INDIA)

Biotechnological Methods of Pollution Control

ISBN 978-93-5111-208-2
© Reserved

All Rights Reserved. No Part of this book may be reproduced in any manner without written permission.

Published in 2014 in India by

RANDOM PUBLICATIONS

4376-A/4B, Gali Murari Lal, Ansari Road
New Delhi-110 002
Phone : +91-11-43580356, +91-11-23289044
e-mail: randomexports@gmail.com, sales@randompublications.com, info@randompublications.com

Reprinted 2026

Type Setting by : Keystoneprintads, Delhi-110051

Preface

This book, one of the first on the subject, describes biotechnological pollution control methods which try to achieve the two major objectives of cost-effectiveness and environmental compatibility. Biotechnology offers the nearly all economical and environmentally kind technique for air pollution control when commerce with the removal of odorous and toxic contaminants from engineering and municipal airstreams. When emit in large amounts, volatile organic compounds and inorganic odorous compounds generate hazards to the ecosystem and healthiness effects to humans.

Bioremediation is a key area of 'white' biotechnology, because the elimination of a wide range of pollutants from water and soils is an absolute requirement for sustainable development. There are numerous processes of cleaning water, industrial effluents and solid wastes, using microorganisms aerobically and anaerobically. Some of them are quite sophisticated, while others are simple and adapted to the conditions of developing countries.

Many industrial processes cause volatile organic and inorganic odorous compounds to be emitted in large quantities. These compounds create hazards to the ecosystem and to humans. The demand for odor and air pollution control systems that provide breathable air is therefore growing. Biotechnology offers one of the most economical and environmentally friendly methods for controlling odor and air pollution.

The way pollution controls are often built into the production process makes any estimation of their cost extremely difficult. In addition, pollution controls often discourage new investment and production, but because the value of what is not produced is not seen, no one currently calculates such indirect costs. The federal government has, however, estimated a subset of costs—namely, direct expenditures on pollution controls. These expenditures cost governments and private entities an estimated $50 billion plus in 2002 alone. Some thirty-one billion dollars was spent on air-pollution abatement, seventeen billion on water-pollution controls, and eight billion for a variety of solid waste, hazardous waste, and other programs. The method of presentation is very clear and lucid, which can be easily followed by the students.

I thank all members of my team who have helped in the preparation of the book. My special thanks go to "Random Publications" who have published the book.

– Anand Prakash

Contents

1

Waste Management and Water Pollution

When toxic substances enter lakes, streams, rivers, oceans, and other water bodies, they get dissolved or lie suspended in water or get deposited on the bed. This results in the pollution of water whereby the quality of the water deteriorates, affecting aquatic ecosystems. Pollutants can also seep down and affect the groundwater deposits. The effects of water pollution are not only devastating to people but also to animals, fish, and birds. Polluted water is unsuitable for drinking, recreation, agriculture, and industry. It diminishes the aesthetic quality of lakes and rivers. More seriously, contaminated water destroys aquatic life and reduces its reproductive ability. Eventually, it is a hazard to human health. Nobody can escape the effects of water pollution.

IMPORTANCE OF WATER

Water is our lifeline that bathes us and feeds us. In ancient cultures water represented the very essence of life. The Romans were the first to pipe water into their growing cities, especially with their aqueducts. They also realised that sewage water could cause damage to their people, and needed to be removed from large areas of people. Water has played a role not only in the history of countries, but in religion, mythology, and art. Water in many religions cleanses the soul through holy water.The water at Lourdes, France is thought by many religions to be sacred water with healing powers. In Egyptian mythology, the Nu was the beginning of everything and represented water. It brought life to their people, but in drought, produced chaos. The water or hydrologic cycle explains interactions between the atmosphere, hydrosphere, and lithosphere. The water or hydrologic cycle is a major driving force on our planet. Water is in constant motion, evapourating into the atmosphere from oceans, lakes, rivers and streams.

When the atmosphere can no longer support the moisture within the clouds, we experience rain, snow, hail, or sleet. Some water is locked in the form of ice at the polar caps and in glaciers. Water melts in the spring, producing runoff, that percolates through the Earth as groundwater (subsurface) or makes its way back to the sea (surface). The oceans contain most of the water, but it is salt water which is unusable by most organisms. Only pure H_2O (water) can interact with organisms. The movement of the oceans also has a direct effect on the atmosphere. The atmosphere is that envelope of gas that keeps organisms living on this planet. Oceans and atmosphere interact to give us weather. Water provides the Earth with the capacity of supporting life. An organism doesn't have to be told how important water is to their existence. An amphibian knows to lay their eggs in water or else there will be no new born. Even flies know to lay their eggs in fresh water.The only organism that doesn't understand the importance of water is humans, especially in industrialized countries.

Children in those societies turn on the water in a sink and never think about the trouble someone has gone for that "miracle" to occur. Dams, reservoirs, filtering plants, and pipes all bring clean water when the facet is turned on. Sewage water is only mixed with recycled water supplies after the water goes through rigorous cleaning methods. Water borne diseases are any illnesses caused by drinking contaminated water. Diseases can include infection from bacteria (Salmonella), viruses, or by small parasites (Cryptosporida, Giardia, and Toxoplasma). These organisms and viruses cause diseases like cholera, typhoid fever, malaria, botulism, polio, dysentery, giardia, and hepatitis. A one of the first symptoms of these diseases is diarrhea, which cause about three million deaths throughout the world.Sewage is sometimes discharged into rivers, where children downstream might be taking a bath or using the water to drink. The simplest treatment method is boiling. Just bring the water to a boil for at least one minute, then allow it to cool. But this is not always effective in heavily chemical polluted water supplies. Without water, organisms could not exist.

Water is a resource that should not be taken for granted. It needs to be conserved, just as we save other valuable resources. Water is one of the weirdest compounds known to humans. The difference between the boiling point and freezing point of water is one of the largest ranges of any compound. It is this span of temperature that mirrors the range of where life can exist, from bacteria to humans. Water also has a very high specific heat, which means that it can absorb or lose much heat before its temperature changes. This is important in maintaining body heat in mammals.

It also takes a lot of energy before vapourization can occur. For this reason, water evapourates slowly from ponds and lakes, where many life forms are dependent on a stable, warm environment. Water is less dense in its solid state than in its liquid state, so that ice floats instead of sinking. This property permits life to develop in polar and subpolar regions where ice floats and allows life to continue living below the surface. If ice were heavier than water, it would sink, and more ice would form on top of it. As a result, all life in the waters would be trapped in the ice in the many areas of the world where it gets cold enough to freeze water.

Resistance of Water to a Disturbance

Water also exhibits viscosity. One can observe the effects of viscosity alongside a stream or river with uniform banks. The water along the banks is nearly still, while the current in the centre may be swift. This resistance between the layers is called viscosity. This property allows smaller fish to live near the shore, while larger fish are able to swim efficiently in strong currents. Viscosity is also responsible for the formation of eddies, creating turbulence that leads to good mixing of air in the water and more uniform distribution of microscopic organisms. Domestic sewage refers to waste water that is discarded from households.

Also referred to as sanitary sewage, such water contains a wide variety of dissolved and suspended impurities. It amounts to a very small fraction of the sewage by weight. But it is large by volume and contains impurities such as organic materials and plant nutrients that tend to rot. The main organic materials are food and vegetable waste, plant nutrient come from chemical soaps, washing powders, etc. Domestic sewage is also very likely to contain disease-causing microbes. Thus, disposal of domestic waste water is a significant technical problem.

Industrial Effluents

Waste water from manufacturing or chemical processes in industries contributes to water pollution. Industrial waste water usually contains specific and readily identifiable chemical compounds. During the last fifty years, the number of industries in India has grown rapidly. But water pollution is concentrated within a few subsectors, mainly in the form of toxic wastes and organic pollutants. Out of this a large portion can be traced to the processing of industrial chemicals and to the food products industry. In fact, a amount of large- and medium-sized industries in the region covered by the Ganga Action Plan do not have adequate effluent treatment facilities. Most of these defaulting industries are sugar mills, distilleries, leather processing industries, and thermal power stations. Most major industries have treatment facilities for industrial effluents. But this is not the case with small-scale industries, which cannot afford enormous investments in pollution control equipment as their profit margin is very slender.

CONCEPTS OF WASTE MANAGEMENT

There are a number of concepts about waste management, which vary in their usage between countries or regions.

The waste hierarchy:

- Reduce
- Reuse
- Recycle Classifies waste management strategies according to their desirability.

The waste hierarchy has taken many forms over the past decade, but the basic concept has remained the cornerstone of most waste minimization strategies. The aim of the waste hierarchy is to extract the maximum practical benefits from products and to generate the minimum amount of waste. Some waste management experts have recently incorporated a 'fourth R': "Re-think", with the implied meaning that the present system may have fundamental flaws, and that a thoroughly effective system of waste management may need an entirely new way of looking at waste. Some "re-think" solutions may be counter-intuitive, such as cutting fabric patterns with slightly more "waste material" left — the now larger scraps are then used for cutting small parts of the pattern, resulting in a decrease in net waste.

This type of solution is by no means limited to the clothing industry. Source reduction involves efforts to reduce hazardous waste and other materials by modifying industrial production. Source reduction methods involve changes in manufacturing technology, raw material inputs, and product formulation. At times, the term "pollution prevention" may refer to source reduction. Another method of source reduction is to increase incentives for recycling.

Many communities in the United States are implementing variable rate pricing for waste disposal which has been effective in reducing the size of the municipal waste stream.Source reduction is typically measured by efficiencies and cutbacks in waste. Toxics use reduction is a more controversial approach to source reduction that targets and measures reductions in the upstream use of toxic materials. Toxics use reduction emphasis's the more preventive aspects of source reduction but, due to its emphasis on toxic chemical inputs, has been opposed more vigourously by chemical manufacturers. Toxics use reduction progammemes have been set up by legislation in some states.

Recovery

A relatively recent idea in waste management has been to treat the waste material as a resource to be exploited, instead of simply a challenge to be managed and disposed of. There are a number of different methods by which resources may be extracted from waste: the materials may be extracted and recycled, or the calorific content of the waste may be converted to electricity. The process of extracting resources or value from waste is variously referred to as secondary resource recovery, recycling, and other terms.

The practice of treating waste materials as a resource is becoming more common, especially in metropolitan areas where space for new landfills is becoming scarcer. There is also a growing acknowledgement that simply disposing of waste materials is unsustainable in the long term, as there is a finite supply of most raw materials. There are a number of methods of recovering resources from waste materials, with new technologies and methods being developed continuously.In some developing nations some resource recovery already takes place by way of manual labourers who sift through un-segregated waste to salvage material that can be sold in the recycling market. These unrecognized workers called waste pickers or rag pickers, are part of the informal sector, but play a significant role in reducing the load on the Municipalities' Solid Waste Management departments. There is an increasing trend in recognizing their contribution to the environment and there are efforts to try and integrate them into the formal waste management systems, which is proven to be both cost effective and also appears to help in urban poverty alleviation. However, the very high human cost of these activities including disease, injury and reduced life expectancy through contact with toxic or infectious materials would not be tolerated in a developed country.

Recycling

Recycling means to recover for other use a material that would otherwise be considered waste. The popular meaning of 'recycling' in most developed countries has come to refer to the widespread collection and reuse of various everyday waste materials. They are collected and sorted into common groups, so that the raw materials from these items can be used again (recycled). In developed countries, the most common consumer items recycled include aluminum beverage cans, steel, food and aerosol cans, HDPE and PET plastic bottles, glass bottles and jars, paperboard cartons, newspapers, magazines, and cardboard. Other types of plastic are also recyclable, although not as commonly collected. These items are usually composed of a single type of material, making them relatively easy to recycle into new products. The recycling of obsolete computers and electronic equipment is important, but more costly due to the separation and extraction problems. Much electronic waste is sent to Asia, where recovery of the gold and copper can cause environmental problems (monitors contain lead and various "heavy metals", such as selenium and cadmium; both are commonly found in electronic items).

Recycled or used materials have to compete in the marketplace with new (virgin) materials. The cost of collecting and sorting the materials often means that they are equally or more expensive than virgin materials. This is most often the case in developed countries where industries producing the raw materials are well-established. Practices such as trash picking can reduce this value further, as choice items are removed (such as aluminum cans).

In some countries, recycling progammemes are subsidized by deposits paid on beverage containers The economics of recycling junked automobiles also depends on the scrap metal market except where recycling is mandated by legislation.However, most economic systems do not account for the benefits to the environment of recycling these materials, compared with extracting virgin materials. It usually requires significantly less energy, water and other resources to recycle materials than to produce new materials

- Recycling 1000 kg of aluminum cans saves approximately 5000 kg of bauxite ore being mined and prevents the generation of 15.17 tons CO_2 eq greenhouse gases
- Recycling steel saves about 95 per cent of the energy used to refine virgin ore

Consumer vs. Machine Waste Separation

In many areas, material for recycling is collected separately from general waste, with dedicated bins and collection vehicles. Other waste management processes recover these materials from general waste streams. This usually results in greater levels of recovery than separate collections of consumer-separated beverage containers, but are more complex and expensive. When consumer-separated recycling is a government requirement, waste is often not well separated due of either ignorance or contempt of the rules.

This results in glass containers that may have metal lids still attached and rotted food inside, aluminum cans full of chewing tobacco spit and cigarette butts, corrugated paper boxes soiled with oils, solvents, or rotting food, and inclusion of incompatible plastic types in a plastics recycling bin. This can all lead to process contamination, work stoppage, a system cleanout, and landfill disposal of the contaminated batch of otherwise recyclable materials. Re-sorting of consumer-separated wastes is often needed to prevent recycling process contamination.

A common method of machine sorting of complex waste streams is to shred the entire stream into a fine particulate of similar size. A magnetic conveyor belt removes ferrous metals from this particulate, and cyclonic separation towers separate objects from the waste stream by mass. Spectral imaging such as with X-rays can further separate glass and various metals from the stream by scanning for x-ray absorption and firing precise puffs of air at the falling pieces to push them sideways into various sorting bins. The remainder of the unsorted shredded material is known as fluff and contains mostly plastics, paper and other organic materials.

When vehicles are shredded and processed in this manner for recycling, often a large mass of fluff results from the plastics used in the seat cushions, dashboard, roof liner, carpeting, and so forth. There are not many well-established processes for further separation and recycling of fluff, other than incineration or pyrolysis.

WASTE MANAGEMENT TECHNIQUES:

Managing municipal waste, industrial waste and commercial waste has traditionally consisted of collection, followed by disposal. Depending upon the type of waste and the area, a level of processing may follow collection. This processing may be to reduce the hazard of the waste, recover material for recycling, produce energy from the waste, or reduce it in volume for more efficient disposal.Collection methods vary widely between different countries and regions, and it would be impossible to describe them all. In Australia most urban domestic households have a 240 litre (63.4 gallon) bin that is emptied weekly by the local council. Many areas, especially those in less developed areas, do not have a formal waste-collection system in place.In Canadian urban centres curbside collection is the most common method of disposal, whereby the city collects waste, and or recyclables, and or organics on a scheduled basis from residential areas.

In rural areas people dispose of their waste at transfer stations. Waste collected is then transported to a regional landfill.Disposal methods also vary widely. In Australia, the most common method of disposal of solid waste is in landfill sites, as it is a large country with a low-density population. By contrast, in Japan it is more common for waste to be incinerated, because the country is smaller and land is scarce.

Landfill

Disposing of waste in a landfill is the most traditional method of waste disposal, and it remains a common practice in most countries. Historically, landfills were often established in disused quarries, mining voids or borrow pits. A properly-designed and well-managed landfill can be a hygienic and relatively inexpensive method of disposing of waste materials in a way that minimizes their impact on the local environment. Older, poorly-designed or poorly-managed landfills can create a number of adverse environmental impacts such as wind-blown litter, attraction of vermin, and generation of leachate which can pollute groundwater and surface water. Another byproduct of landfills is landfill gas (mostly composed of methane and carbon dioxide), which is produced as organic waste breaks down anaerobically.

This gas can create odour problems, kill surface vegetation, and is a greenhouse gas. Design characteristics of a modern landfill include methods to contain leachate, such as clay or plastic lining material. Disposed waste is normally compacted to increase its density and stabilize the new landform, and covered to prevent attracting vermin (such as mice or rats) and reduce the amount of wind-blown litter. Many landfills also have a landfill gas extraction system installed after closure to extract the landfill gas generated by the decomposing waste materials.

Gas is pumped out of the landfill using perforated pipes and flared off or burnt in a gas engine to generate electricity. Even flaring the gas is a better environmental outcome than allowing it to escape to the atmosphere, as this consumes the methane, which is a far more potent greenhouse gas than carbon dioxide. Many local authorities, especially in urban areas, have found it difficult to establish new landfills due to opposition from owners of adjacent land. Few people want a landfill in their local neighbourhood. As a result, solid waste disposal in these areas has become more expensive as material must be transported further away for disposal (or managed by other methods) This fact, as well as growing concern about the impacts of excessive materials consumption, has given rise to efforts to minimize the amount of waste sent to landfill in many areas.

These efforts include taxing or levying waste sent to landfill, recycling the materials, converting material to energy, designing products that use less material, and legislation mandating that manufacturers become responsible for disposal costs of products or packaging. A related subject is that of industrial ecology, where the material flows between industries is studied. The by-products of one industry may be a useful commodity to another, leading to a reduced materials waste stream.

Some futurists have speculated that landfills may one day be mined: as some resources become more scarce, they will become valuable enough that it would be economical to 'mine' them from landfills where these materials were previously discarded as valueless. A related idea is the establishment of a 'mono-fill' landfill containing only one waste type as a method of long-term storage.

Incineration

Incineration is a waste disposal method that involves the combustion of waste at high temperatures. Incineration and other high temperature waste treatment systems are described as "thermal treatment". In effect, incineration of waste materials converts the waste into heat, gaseous emissions, and residual solid ash. Other types of thermal treatment include pyrolysis and gasification. A waste-to-energy plant (WtE) is a modern term for an incinerator that burns wastes in high-efficiency furnace/boilers to produce steam and/or electricity and incorporates modern air pollution control systems and continuous emissions monitors. This type of incinerator is sometimes called an energy-from-waste (EfW) facility. Incineration is popular in countries such as Japan where land is a scarce resource, as they do not consume as much area as a landfill. Sweden has been a leader in using the energy generated from incineration over the past 20 years. Denmark also extensively uses waste-to-energy incineration in localized combined heat and power facilities supporting district heating plans. Incineration is carried out both on a small scale by individuals, and on a large scale by industry.

It is recognized as a practical method of disposing of certain hazardous waste materials (such as biological medical waste), though it remains a controversial method of waste disposal in many places due to issues such as emission of gaseous pollutants.Breaking down complex chemical chains such as dioxin through the application of heat usually cannot be done by simply burning the material at the temperatures seen in an open-air fire. It is often necessary to supplement the combustion process with gas or oil burners and air blowers to raise the temperature high enough to result in molecular breakdown. Alternately, the exhaust gases from an natural air fire may pass through tubes heated to sufficiently high temperatures to trigger thermal breakdown. Thermal breakdown of pollutant molecules can indirectly create other pollution problems. Dioxin breakdown begins at 1000°C, but at the same time poisonous nitrogen oxides and ozone begin to form when atmospheric nitrogen and oxygen break down at 1600°C. This undesired oxide formation may require further catalytic treatment of the exhaust gases.

Composting and Anaerobic Digestion

Waste materials that are organic in nature, such as plant material, food scraps, and paper products, are increasingly being recycled. These materials are put through a composting and/or digestion system to control the biological process to decompose the organic matter and kill pathogens. The resulting stabilized organic material is then recycled as mulch or compost for agricultural or landscaping purposes.

There are a large variety of composting and digestion methods and technologies, varying in complexity from simple windrow composting of shredded plant material, to automated enclosed-vessel digestion of mixed domestic waste. These methods of biological decomposition are differentiated as being aerobic in composting methods or anaerobic in digestion methods, although hybrids of the two methods also exist.Examples:The Green Bin Progammeme, a form of organic recycling used in Toronto, Ontario and surrounding municipalities including Markham, Ontario, Canada, makes use of anaerobic digestion to reduce the amount of garbage shipped to Michigan, in the United States.

This is the newest facet of the 3-stream waste management system has been implemented in the city and is another step towards the goal of diverting 70 per cent of current waste away from the landfills. Green Bins allow any organic waste that in the past would have formed landfill waste to be composted and turned into nutrient rich soil. Examples of waste products for the Green Bin are food products and scraps, soiled papers and sanitary napkins. The Green Bin Progammeme, a form of organic recycling used in Toronto, Ontario and surrounding municipalities including Markham, Ontario, Canada, makes use of anaerobic digestion to reduce the amount of garbage shipped to Michigan, in the United States.

This is the newest facet of the 3-stream waste management system has been implemented in the city and is another step towards the goal of diverting 70 per cent of current waste away from the landfills. Green Bins allow any organic waste that in the past would have formed landfill waste to be composted and turned into nutrient rich soil. Examples of waste products for the Green Bin are food products and scraps, soiled papers and sanitary napkins. Currently Markham, like the other municipalities in the The Green Bin Progammeme is currently being studied by other Municipalities in the province of Ontario as a way of diverting waste away from the landfills. Notably, Toronto and Ottawa are in the preliminary stages of adopting a similar progammeme.The City of Edmonton, Alberta, Canada has adopted large-scale composting to deal with its urban waste. Its composting facility is the largest of its type in the world, representing 35 per cent of Canada's centralized composting capacity. The $100 million co-composter and various recycling progammemes enable Edmonton to recycle 60 per cent of its residential waste. The co-composter itself is 38,690 square meters in size, equivalent to 8 football fields. It's designed to process 200,000 tons of residential solid waste per year and 22,500 dry tons of bio-solids, turning them into 80,000 tons of compost annually.

Mechanical Biological Treatment

Mechanical biological treatment (MBT) is a technology category for combinations of mechanical sorting and biological treatment of the organic fraction of municipal waste. MBT is also sometimes termed BMT- Biological Mechanical Treatment- however this simply refers to the order of processing. The "mechanical" element is usually a bulk handling mechanical sorting stage. This either removes recyclable elements from a mixed waste stream (such as metals, plastics and glass) or processes it in a given way to produce a high calorific fuel given the term refuse derived fuel (RDF) that can be used in cement kilns or power plants.

Systems which are configured to produce RDF include Herhofand Ecodeco. It is a common misconception that all MBT processes produce RDF. This is not the case. Some systems such as ArrowBio simply recover the recyclable elements of the waste in a form that can be sent for recycling. The "biological" element refers to either anaerobic digestion or composting. Anaerobic digestion breaks down the biodegradable component of the waste to produce biogas and soil conditioner.

The biogas can be used to generate renewable energy. More advanced processes such as the ArrowBio Process enable high rates of gas and green energy production without the production of RDF. This is facilitated by processing the waste in water. Biological can also refer to a composting stage. Here the organic component is treated with aerobic microorganisms. They break down the waste into carbon dioxide and compost. There is no green energy produced by systems simply employing composting. MBT is gaining increased recognition in countries with changing waste management markets such as the UK and Australia where WSN Environmental Solutions has taken a leading role in developing MBT plants

Pyrolysis and Gasification

Pyrolysis and gasification are two related forms of thermal treatment where waste materials are heated to high temperatures with limited oxygen availability. The process typically occurs in a sealed vessel under high pressure. Converting material to energy this way is more efficient than direct incineration, with more energy able to be recovered and used. Pyrolysis of solid waste converts the material into solid, liquid and gas products.

The liquid oil and gas can be burnt to produce energy or refined into other products. The solid residue (char) can be further refined into products such as activated carbon. Gasification is used to convert organic materials directly into a synthetic gas composed of carbon monoxide and hydrogen. The gas is then burnt to produce electricity and steam. Gasification is used in biomass power stations to produce renewable energy and heat.

RADIOACTIVE WASTE MANAGEMENT

FRAMING THE PROBLEM

The United States is at a "gridlock" position regarding nuclear waste management. Existing nuclear power plants, left to manage wastes in the absence of a coherent national policy, have become de facto long-term storage sites, using facilities designed only to temporarily house such materials.

Radioactive waste has emerged as one of the issues inhibiting further development of the nuclear power industry, and the safety implications of forcing every power plant to handle wastes on a longer-term basis are severe.Former nuclear weapons production sites face even more significant problems with radioactive waste management. The scale and scope of the cleanup at these sites is enormous; officials estimate that seventy-five years and $300 billion will be required to remediate these facilities. The Department of Energy (DOE), which is responsible for these sites, faces both an environmental and an administrative quagmire as it attempts to clean up after fifty years of nuclear weapons production. The lesson that follows is an introduction to these issues, focusing on an understanding of how, over time, the problems associated with radioactive waste have developed. This lesson is predicated on the assumption that radioactive waste management is not a single task to be accomplished but is rather a multidimensional issue that needs to be understood and addressed physically, socially, and historically.

The multiple aspects of radioactive waste management this lesson examines include physical and value-oriented issues. What radioactive waste is, where it is, how it has been generated and handled, what some of the difficulties are in handling, treating, and disposing of the stuff—these are central issues in radioactive waste managment. Values are also critical in understanding radioactive waste, and must be understood in their social and historical contexts—how radioactive waste has been understood (or misunderstood, as the case may be), how it has meant different things to different people, and how these meanings have changed over time. An understanding of these value-oriented issues will enable a discussion of the political problem of radioactive waste, how it has developed into a metaphorical "hot potato" that nobody is willing to hold. Understand both values and the physical aspects of radioactive waste will help foster an understanding of the bureaucratic problems involved in managing this material —resolving the radioactive waste problem will involve administrative and managerial feats that the DOE has not yet shown itself capable of handling.

The radioactive waste "problem" is not a new one; health and safety concerns have been associated with radioactive materials throughout the twentieth century. Indeed, such concerns have existed almost from the discovery of radioactivity in 1895, when a German physicist, Willem Roentgen, identified what he called "x-rays," and developed a technique for producing them. The medical community quickly adopted x-rays as a useful diagnostic tool; within a year, x-ray machines could be found in every major city.

In early 1890's identified a new element, radium, that had radioactive properties. Radium, like x-rays, was adopted for several industrial uses. Observers soon noted that these radioactive materials were associated with harmful side effects. Medical x-ray workers and many scientists were suffering from skin burns, blood disorders, and a variety of otherwise rare cancers; indeed, Marie Curie suffered terribly from bone diseases stemming from her prolonged exposure to radium and other radioactive materials. The women who worked as watch-dial painters began to develop horrible forms of jaw and throat problems. Health workers learned that these women would use their mouths to "point" their brushes, thereby ingesting what often amounted to lethal quantities of radioactive compounds.Several interested organizations and individuals moved, in response to the recognition of the health effects of radioactive materials, to form independent bodies to study the issues and set voluntary standards for exposure.

The Manhattan Project: Large-Scale Work with Radioactive Materials

Industrial work with radioactive materials was relatively small in scale through the 1930s, but events during and after World War II changed matters dramatically. Indeed, the work done to construct atomic weapons, begun as war raged across Europe and the Pacific, continues to have profound environmental and political effects in the contemporary world.

The Decision to Build the Bomb

Efforts to construct an atomic bomb in the United States began in the late 1930s, under the assumption that the Allies were in a race with Germany. In 1938, two German physicists announced that they had demonstrated fission—the splitting of the atom. Scientists recognized that a controlled fission reaction could be made into a new, incredibly destructive weapon. Many of the most eminent physicists in the. were emigres from Germany, Austria, Hungary, and Italy; having experienced fascism, they were alarmed by the prospect of an atomic weapon in the hands of Hitler's regime. The efforts of these emigre scientists were crucial in prompting the American government to undertake what became known as the Manhattan Project.

The Manhattan Project refers to the efforts of the Manhattan Engineering District (MED), the organization created by the U.S. Army to carry out the atomic bomb development progammeme. This was truly an enormous industrial and scientific undertaking, and was the largest construction project ever undertaken at the time.

Health and Safety in the Manhattan Engineering District

Radiation safety was a priority in the MED. Personnel protection was strongly emphasized, even more than in other types of production work, and the NCRP safety guidelines proved quite effective in this regard. Scientists were in short supply and even enlisted men (who died by the thousands overseas) were well protected. The unique nature of the research and production work made trained personnel even more valuable than normal; there simply were no replacements available. Management thus emphasized health and safety precautions, as the project could not afford to lose trained personnel to radiation exposure. Site selection reflected safety concerns of a different sort.. The MED chose to locate its primary plutonium production facility (Hanford) in eastern Washington, rather than nearer to the labouratories in Chicago or Oak Ridge, because of fears of the possible consequences of a major accident.

The Hanford facility utilized new and untested production techniques with extremely dangerous substances. Managers did not know exactly what the risks of operating such a plant were, but they assumed that they were significant. Thus, they chose to place the facility in a remote location so that if a catastrophic accident did occur it would affect relatively few people. Waste management was not a high priority during the war; this is not a surprise given the pressures of defeating Germany and Japan. Although managers clearly understood that production processes would generate vast quantities of highly toxic materials, especially from the separations processes at Hanford, they did not give much thought to the development of a long-term waste disposal plan. The MED sought to minimize the immediate risks associated with these wastes, but deliberately deferred developing disposal solutions for the various waste management problems until after the war. Even though waste management and health concerns were re-evaluated after the war ended, the MED never made the solution of waste problems a priority. As it turned out, neither did its successor, the Atomic Energy Commission.

The Legacy of the MED and the Rise of the Cold War

The MED was ultimately successful in developing atomic bombs, two of which were dropped on Japan in August, 1945. Atomic weapons became a central element of American diplomacy from the moment of their first use, and "the bomb" played a critical role in the emerging Cold War between the US and the USSR. The tensions between East and West emerged even before the war ended and escalated throughout the 1940s and 1950s. As a result, with the exception of a brief post-war lull, the American nuclear weapons production effort begun during the war continued at a high level well into the 1960s.The rapid transition from WWII to the Cold War helps to explain the lack of attention paid by the Manhattan Engineering District and the Atomic Energy Commission to waste managment issues. As atomic weapons became a defining symbolic and literal expression of American power, the need to continue production at all costs continued. There was no extended post-war shut-down of the wartime production network, no time to develop new approaches to waste production and management. The pressure of the Cold War made for a very uneasy peace indeed.

EFFECTS OF WATER POLLUTION

The effects of water pollution are not only devastating to people but also to animals, fish, and birds. Polluted water is unsuitable for drinking, recreation, agriculture, and industry. It diminishes the aesthetic quality of lakes and rivers. More seriously, contaminated water destroys aquatic life and reduces its reproductive ability. Eventually, it is a hazard to human health. Nobody can escape the effects of water pollution.

Effects of Run-off Pollution

Rain picks up dirt and silt and carries it into the water. If the dirt and silt settle in the water body, then these sediments prevent sunlight from reaching aquatic plants. If the Sun can't reach the plants, these perish. These sediments also clog fish gills and smother organisms that live on the bottom of the body of the water.

Effects of Oil Pollution and Antifreeze

If oil is spilled on the water, the effects on the ecosystem and the components are harmful. Many animals can be annihilated in case they ingest oil. Oil contaminated prey may be a reason of death for many. If the oil coats the feathers of birds, these may die. Oil and antifreeze makes the water have a foul odour and there is a sticky film on the surface of water that kills animals. Oil is the most harmful pollutant in the water.

Contaminated Ground Water Effects

If contaminated water enters the ground, there may be serious effects. People may become very sick and there is a probability of developing liver or kidney problems and cancer or other illnesses.

Fertilizers and Other Chemicals

Nitrates in drinking water leads to diseases of infants that may lead to their death. Cadmium is a metal in sludge-derived fertilizer. This can be absorbed by crops. When people ingest this, they may cause diarrheal disorders, liver and kidney damage. The inorganic substances like mercury, arsenic and lead are the causes of pollution. Other chemicals can also lead to problems concerning the taste, smell and colour of water. Pesticides, PCBs and PCPs are all poisonous to all sorts of life. Pesticides are used in farming, homes and forestry. PCBs are found as insulators in old electrical transformers. PCPs are found in products like wood preservatives.

Effects of Agricultural Water Pollution

Rain and irrigation water drains off cultivated land that has been fertilized and treated with pesticides, the excess nitrogen and poisons are mixed with it into the water supply. These pesticides are toxic and pollute the water in a different mode. Aquatic plants growth cause deoxyge-nation of water and annihilate flora and fauna in a stream, lake and river. Fertilizers enhance the growth of bacteria that are in water and increase the concentration of bacteria to hazardous levels.

Effects of Thermal Water Pollution

Machinery in the industries are cooled with water from lakes and rivers. This water reaches the river in a heated state. This water decreases the ability of the aquatic system to hold oxygen and raises the growth of warm water species.

Effects of Heavy Metal Water Pollution

Heavy metals like lead, mercury, iron, cadmium, aluminum and magnesium are present in water sources. If these metals are present in the sediment, these reach the food chain through plants and aquatic animals. This causes heavy metal poisoning in case the level in the water is very high.

DOMESTIC WASTE MANAGEMENT STRATEGIES

GENERAL CONSIDERATIONS

Domestic wastes in park and tourism developments come from toilets and urinals, showers, bathroom sinks, kitchen sinks, laundry facilities, and most floor drains. Varying amounts of water are used to carry these wastes through pipes to a treatment point. Different processing techniques are used to convert the wastes for reuse or disposal. True sustainable development would not permit direct disposal of either liquids or solids before reuse. Responsible waste management recognizes the value of reducing water needs. Properly treated wastewater can be used for toilet flushing (after approved disinfection) and irrigation (landscape needs or agricultural plantings). The solids separated from domestic waste may be incorporated into the solid wastes (garbage) and composted into an excellent soil amendment product or anaerobically digested to produce a gaseous energy source and a residue that can also be used as a soil amendment. If wastewater recycling for toilet flushing is used without concurrent irrigation of vegetation, there will be a volume of excess liquid that must be sensitively disposed of. This will be the amount of liquid wastes coming in each day from all sources, less the recirculated amount used for toilet flushing. Best disposal practice would use a suitable area for subsurface movement of the liquid through soils that provide good filtering and additional treatment before reaching any body of water. Direct discharge to a waterway should be the lowest priority of alternatives investigated.

Using treated, recycled wastewater for toilet flushing, rather than an equal volume of drinking-quality water, would save a large amount of water. There are also strong economic and environmental reasons for using one of more than 20 different low-volume flush toilets currently on the market, rather than the more conventional water-wasting models. In seawater development sites, toilets may be flushed with seawater and processed with septic tank and subsurface treatment and disposal. This provides a reliable system if the corrosive nature of seawater is addressed in selection of materials used in the piping, plumbing, pumping, and treatment systems. Similar water conservation can be realised through the use of flow-restricted, fine-spray shower heads, and sink faucet aerators. Spring-loaded handles should be installed on faucets. Laundry waste volume can be reduced through the use of machines with suds-saver cycles. Park and tourism developments that make use of water conservation devices should make an effort to educate their visitors on the amount of water (or energy) they have conserved during their stay, compared with the conventional development. It is important to monitor the conservation devices in order to evaluate their performance. As with all pollution, sewage is best treated by not creating it in the first place. Human waste only becomes waste when there is no use for it.

Specific Waste Treatment Technologies

Conditions at sustainable developments favour simple, reliable, passive techniques for waste treatment. Systems that have few moving parts, controls, and monitoring requirements should be used. Continuity of operation and maintenance knowledge is essential. Waste treatment systems must also be able to withstand seasonal fluctuations in use. Preparation for an idle period should be simple, and no energy should be required when the system is not being used. If an innovative technology is to be employed for waste treatment, it should have a conventional backup system, or be sufficiently redundant that multiple failures are unlikely.

Dry Toilets

A composting toilet consists of a large tank located directly below the toilet room. Wastes enter the tank through a large diameter chute connecting to the toilet, and decompose in an oxygen rich environment. No water is used for the toilet, but a bulking agent (such as wood shavings) is added to improve liquid drainage and aeration, and to provide fuel. A small fan draws air through the tank and up the vent pipe to ensure adequate oxygen for decomposition and odorless operation.

Internal components (such as ducts, baffles, and rotating tines) enhance the composting process. The finished compost can be removed from the lower end of the tank about once each year. It can be used as a fertilized for soil. Composting toilets need a mild temperature, moisture, fuel, and air to function. Liquid may have to be added to the tank to keep the compost pile moist during periods of little use or a bulking agent added periodically to improve the compost texture. Composting toilets have several advantages over other systems - *e.g.*, no water is used and only a small amount of energy is needed for an exhaust fan; valuable nutrients are used to benefit soils; proper maintenance requires little time.

There are some disadvantages to this system. The user is close to the sewage treatment systems and this bothers some visitors. Without proper maintenance, the tank can become anaerobic and unpleasant odors arise. Undesirable pests can take up residence in the composting tank. The availability of a bulking agent in some areas can also be a problem.

Anaerobic Waste Treatment

Anaerobic waste treatment (septic systems) is accomplished through microorganisms living in the wastewater. Anaerobic microorganisms work in an environment where there is no free oxygen. Complex reactions and interactions take place with the resulting generation of some offensive gaseous by-products. These unpleasant odors are actually an indication of an efficient progression of the anaerobic process to remove the pollutants from the waste stream. This type of system is reasonable to consider for smaller developments, where the by-product gases can be separated from occupancy areas, dissipated with good air movement, or neutralized with soil or carbon filters. Since slow treatment means longer holding periods (shallow depth tanks), large, isolated treatment and disposal areas are needed.

Treated wastewater (effluent) is usually disposed of in an underground system that passes the effluent through carefully selected undisturbed soil profiles. These soils must further filter and remove nutrients as the effluent makes its way to the groundwater or other bodies of water. One variation of this type of treatment uses part of its treated effluent for toilet flushing. The stored recycled effluent should receive some aeration to ensure odour-free recycle water in the toilets. This type of waste treatment system has several advantages. It is relatively inexpensive to install, and is not complex to operate and maintain. It provides a consistently good quality effluent. Most components can be installed aboveground, or they can be buried or partly buried. Good quality flushing water can be provided to reduce water supply needs and final disposal volumes. The disadvantages are that septic tank odors must be dealt with properly. Several pumps and blowers can be involved, which creates maintenance and power costs, and can cause noise in some developments.

Septic tank solids require proper disposal. An aerobic digestion tank would reduce volumes for disposal. A sand drying bed can be used for digested sludge drying, and the dried sludge should be buried in a landfill. A fairly large area is required for installation. There is no known manufacturer of a complete package system. Prior to committing time and resources, it is prudent to contact actual users of similar systems - talk with a designer, system operator, an owner, and possibly a regulatory agency inspector who has observed performance.

Aerobic Waste Treatment

Aerobic waste treatment is also accomplished with micro-organisms. However in this system, air is bubbled into the treatment process to ensure plenty of free oxygen. Aerobic organisms work about 20 times faster than anaerobic organisms. No offensive gas odors are released in this treatment process. Since they process is so much faster, much less holding time is required and less treatment area is needed. This type of treatment plant is reasonable to consider for all sizes of park and tourism development, and because offensive odors are not normally produced, this type of system can be located close to occupied areas.

The normal effluent quality of a properly designed and operated system meets most all secondary treatment standards With a reasonably small filter, high quality effluent can be produced for irrigation and recycled toilet flushing water. With carefully planned effluent recycling and irrigation reuse, little or no effluent disposal would be required. If disposal is required, several options are possible. There are several manufacturers of small aerobic treatment plants. The best plants for harsh environments and isolated areas are fabricated from durable, low maintenance materials, simple to operate and maintain, designed to use the fewest moving parts, and consistent in effluent quality.

They are also quick to install and put in service, proven in similar installations, and completely assembled and warranted by one manufacturer. A good manufacturer also provides, at a reasonable cost, technical and monitoring services on the new plant for at least one year after initial startup. Advantages of such a system are that all materials in contact with liquids are noncorrosive (plastic, fibreglass, rubber, stainless steel) and can be buried or installed aboveground. The blower is the only moving part and requires only simple, routine maintenance. Effluent quality is consistently good. Excellent recycle and reuse options are available. The process easily handles surge loadings, as well as underloaded periods. Land area required is about 0.1 acre per 10,000 gallons of rated capacity. Site installation and startup can usually be accomplished in five working days or less. Few spare parts are needed. Purchase costs are competitive with other similar-sized treatment units. Unskilled operators have successfully followed infrequent, telephoned system adjustment instructions to maintain top performance.

The disadvantages are that most components are fabricated, assembled, and shipped from Colourado - few local materials would be used. The blower, although small, needs to be run continuously. The noise would have to be muffled, and a constant power cost would be involved. The National Park Service has five of these plants in service in four widely varying environmental areas. One plant has been in service for several years.

Alternative Disinfection Methods

Traditionally, water from conventional treatment systems is disinfected with chlorine or chlorine compounds before being released back into the environment or reused. A side effect of this is that the chlorine or chlorine compounds are very reactive and sometimes produce highly persistent, toxic chemicals. Many environmentalists believe that there is no justification for use of chlorine and its compounds for disinfection. Most public health codes call for disinfection with chlorine, and they would have to be changed to allow for either no disinfection or the use of other disinfectants.

Less desirable possibilities include using less chlorine or removing the chlorine after the proper contact time. Dechlorination requires additional chemical feed and training. The purpose of disinfection is to ensure that no virulent organisms are present after the water has been processed in one of the systems described earlier. The most common alternative disinfectants are ozone and ultraviolet light. An entirely new treatment technology, introduced in September 1993, appears to provide excellent disinfection without the formation of environmentally harmful by-products. The National Park Service is arranging pilot studies at existing water and sewage treatment facilities to evaluate this emerging innovative technology.

THE WASTE PROBLEM IN THE INDUSTRIAL SOCIETY

The waste management system has been defined as the collection of economic activities that accepts solid or liquid waste inputs from households, and/or industry and either recycles or disposes of them. This has been called an "industry", and (in recent years) even a "growth industry". There are a number of large national and some multinational firms that operate profitably in this field. But it is important to recognize from the outset that, while growth in this sector adds to GDP, it is not an indicator of increasing human welfare.

Market Failure

The need for a separate waste management system outside and distinct from both the productive part of the economy (the "factory") and the consumption unit (the "household") is a reflection of market failure in the sense that neither producers nor consumers normally pay for waste management services for their own sake. The fact that households normally pay a small charge for refuse collection is a minor exception to this rule. Despite lip-service to environmental concerns, most waste generators still act and think as though the environment is a free waste-disposal sink. The fact that rivers and lakes, landfills and even the atmosphere and the oceans are finite, and are becoming over-used, has not been translated into a market-determined scarcity cost for environmental services, as would normally be the case for a scarce resource. The reason this has not happened is that the resources in question – except for landfills – are common property "public goods" that cannot be monopolized or "owned" by any individual.

This is the essential market-failure. To dispose of wastes without overloading the environment, it has been necessary to invent a "waste management" sector to provide waste disposal services. In the case of households and communities, these services are paid for by public sector agencies, which are reimbursed through refuse collection fees, water charges and/or property taxes. In the case of industry some waste management services are paid directly (by industry) to comply with regulations imposed and enforced by the public authorities. There is only a weak and imperfect link between the quantity and type of waste generated and the charges for treatment and disposal. In many cases, the costs attributed to polluters are too low to induce significant changes in their behaviour. In many cases, too, the level of treatment or amelioration is inadequate and the resulting environmental damages are accumulating. In short, there is no competitive free market between buyers and sellers of environmental services, as such. Instead, the benefits of waste management services are a rather diffuse common good.

This good is enjoyed equally by all, but it is not paid for in proportion by those who generate the wastes. The intervention by public authorities is an inefficient "second best" solution, relying more on rules and regulations and less on economic in centives than would be optimal. Statistics of the waste management "sector" must be treated with care; they can easily be misinterpreted. (For instance, statistics showing an increase in the aggregate quantity of solid and liquid wastes treated may or may not imply that the quantity of wastes being generated is growing. By the same token, they may or may not imply that the overall pressure on the environment is being reduced. Such a trend may simply be a measure of increasing regulation. Or it could reflect increasing separation between the various stages of production and processing, and of consumption, *i.e.* consequences of specialization, centralization and scale). So, although statistics of the waste management sector have value, on their own they are not likely to be optimal measures of distance from or approach to long-term sustainability.

Delimitation of the Sector in the SIP Project and Overlaps with the Other SIPs

We have already defined the waste management system as the collection of economic activities that receives solid or liquid waste inputs from households, industry and/or other economic sectors and either recycles or disposes of them. For statistical purposes it is useful to represent the sector as any other industrial system: a set of processes characterized by an input - output structure. In this specific case, the process inputs are the different wastes to be treated, whereas the outputs are constituted by the relatively inert residuals of the treatment processes, air and water emissions.

In order to provide an homogeneous representation, a similar input - output structure should be applied to all the six economic sectors of the SIP project, including Tourism and Transport, even if the latter do not produce physical products as outputs. Such a difficulty may be overcome simply by considering that these two sectors provide a service instead of a physical good. The general view of the economic activities of each SIP that results from this general input - output representatio. The figure also puts a special focus on the major waste streams produced by these sectors and accepted by the waste management system. Given the complex interlinks occurring between the waste management system and the other economic sectors, it is important to define appropriate system boundaries among the sectors.

However, the identification of these boundaries is strongly dependent on how waste management is defined. We have adopted the definition of waste management as a sector providing services of waste collection, treatment and disposal for the various economic activities and municipalities. However, depending on the industrial structure in Member States, the management of waste may also take place as a sub-activity of each economic activity. Thus, waste management could also be regarded as an industrial operation or a functional unit occurring in all the economic activities. This would also be in accordance with the concept of statistical unit Functional units such as waste treatment facilities in an establishment or an enterprise can be considered as a Unit of Homogenous Production.

Indeed, the functional role rather than the sectoral one may be the most important for the overall waste management sector in a Member State. The issue is particularly important for the case of industrial waste, because a large number of manufacturing industries provide on-site treatments before discarding their residuals for final disposal. Depending on the perspective adopted (waste management as an independent sector or as a functional unit), the environmental pressure resulting from these on-site treatments could be allocated either to industry (where the treatments take place) or to the waste management sector (because, after all, even on-site treatments are operations of "waste management"). If we adopted a classification system based on the second principle (aggregation of homogeneous processes), the pressure of the on-site treatments should be allocated to the waste management sector. However, for purposes of statistical accounting such an approach would present some problems of data collection.

It is more difficult to gather precise data on the intermediate products of a production unit than to obtain overall input - output Therefore, it seems more convenient to adopt the mill gate as a conceptual boundary between the two sectors and to allocate to the primary waste generators the pressure of the on-site treatments. The waste management sector will be accounted only for the environmental pressure generated by the treatment and final disposal of wastes which leave the mill gate of primary waste generators. That is, those wastes not further treated or recycled by industry and consumers.

System Boundaries between Industry and Waste Management

As a consequence, the manufacturing sector will be accounted both for the pressure generated by on-site treatments of primary wastes (in principle contributing to all the policy fields except "waste") and for the pressure related to the generation of wastes which leave the mill gate (both treated and untreated). It is only the latter pressure that contributes to the policy field "waste". On the other hand the waste management sector will be accounted for the pressure due to off-site treatments (again contributing to all the policy fields except "waste") and for the pressure of final wastes to be disposed of (only contributing to the policy field "waste").

The Waste Management Sector and the Environmental Policy Fields

In principle, the waste management sector produces environmental pressure on the ten policy fields according to the theoretical plan presented. A presentation of the specific environmental pressure caused by the sector is presented. Some general aspects are summarized relating to the major environmental concerns and their policy implications in the context of waste management.

Being a collection of many different pressures arising from great number of waste materials and waste management systems or processes that accepts, treats and disposes of a large number of substances of a various nature, the pressures of the waste management sector may be generally described by means of pressure characterization and a twofold representation of underlying factors. A first order description of both the different waste streams and waste management systems. A general view of the major material and energy flows generated by the various economic activities and connected to the sector is also presented

PRESSURE CHARACTERISTICS

As a general remark, major environmental concerns about waste generation relate both to the problem of resource depletion and to that of environmental pollution Resource aspects include material efficiency and resource intensity for the main production processes from extraction of natural resources, manufacturing, to consumption and final disposal of products. Specific considerations cover:

- Waste prevention, or "more of less" principle;
- Increase of product/material life time, increase of separability, compatibility and disassembly of materials and components;
- Improvements of product efficiency with respect to material and energy consumption;
- Reuse and recycling of waste materials (this practice, instead of new material modifications and/or resource extraction will retain the existing material structure of waste and minimize the additional extraction of virgin material and energy resources for the production of new products. The more cumulative resource inputs are associated to waste material the less sustainable their disposal will be. With respect to pollution, the waste management sector impacts on the natural environment by discarding treatment residuals to air, water bodies or soil.

As a consequence, several objectives can be identified such as:

- To close material cycles; minimization of loading
- To prevent and minimize generation of hazardous waste
- To manage waste collection, handling, treatment and final disposal
- To increase reuse and recycling in order to minimize emissions from waste incineration installations
- Proximity principle: to limit transportation of waste
- To control environmental pollution preferably at the sources of waste.

The long-term objectives of the EU waste policies is to conserve raw materials and energy as well as to ensure the high level of environmental protection. The cause-effect relationships of these two types of pressures are not always recognized. However the evaluation of the overall environmental impacts of waste already includes quantitative (the amount of waste generated), and qualitative aspects (their hazardousness).

Major Waste Streams

An example of the large number of different waste streams produced by the most important economical activities is presented, together with some additional information regarding possible treatment processes

Management Systems

In principle, we note four major economic processes, plus an additional external non-economic sector to include all the substances which are directly discharged into the environment without any special treatment (mainly agriculture residuals). Thus, the processes for the management of solid and liquid wastes can be subdivided as follows:

- Landfilling or land application
- Incineration
- Treatment of liquid wastes, including sewage
- Treatment of special and hazardous wastes
- Direct disposal without treatment

The solid components of materials treated in the second, third and fourth sub sectors usually ends up in landfills for a definitive final disposal. Thus the soil can be regarded as the final "sink" for most wastes (except for the soluble residuals discharged to water courses and the airborne emissions from incineration).

WASTEWATER TREATMENT

Wastewater treatment uses microbes to decompose organic matter in sewage. If too much untreated sewage or other organic matter is added to a lake or stream, dissolved oxygen levels will drop too low to support sensitive species of fish and other aquatic life. Wastewater treatment systems are designed to digest much of the organic matter before the wastewater is released so that this will not occur. Treatment systems use physical, chemical, and biological processes:

- Primary treatment physically removes large solids using grates, screens, and settling tanks.
- Secondary treatment promotes growth of bacteria and other microbes that break down the organic wastes. These biodegradation processes also take place in streams, lakes, and oceans, but the purification systems in nature can easily be overloaded with input of too much organic waste. Secondary treatment prevents this type of pollution by degrading most of the organic matter before the water is released into the environment.
- Tertiary treatment is used only where it is needed to protect the receiving waters from excess nutrients. In tertiary treatment, the concentrations of phosphorus or nitrogen are reduced through biological or chemical processes.
- Disinfection kills disease-causing organisms, most commonly

through chlorination. Sludge, the collection of solids that are removed during wastewater treatment, requires processing to reduce odour and water content. Depending on the disposal method, the sludge also may undergo treatment to decompose organic matter or kill disease-causing organisms.

Wastewater Treatment Principles

Sewage is the wastewater released by residences, businesses and industries in a community. It is 99.94 per cent water, with only 0.06 per cent of the wastewater dissolved and suspended solid material. The cloudiness of sewage is caused by suspended particles which in untreated sewage ranges from 100 to 350 mg/l. A measure of the strength of the wastewater is biochemical oxygen demand, or BOD5. The BOD5 measures the amount of oxygen microorganisms require in five days to break down sewage. Untreated sewage has a BOD5 ranging from 100 mg/l to 300 mg/l. Pathogens or disease-causing organisms are present in sewage. Coliform bacteria are used as an indicator of disease-causing organisms. Sewage also contains nutrients (such as ammonia and phosphorus), minerals, and metals. Ammonia can range from 12 to 50 mg/l and phosphorus can range from 6 to 20 mg/l in untreated sewage.

Sewage treatment is a multi-stage process to renovate wastewater before it reenters a body of water, is applied to the land or is reused. The goal is to reduce or remove organic matter, solids, nutrients, disease-causing organisms and other pollutants from wastewater. Each receiving body of water has limits to the amount of pollutants it can receive without degradation. Therefore, each sewage treatment plant must hold a permit listing the allowable levels of BOD5, suspended solids, coliform bacteria and other pollutants. The discharge permits are called NPDES permits which stands for the National Pollutant Discharge Elimination System.

Preliminary Treatment

Preliminary treatment to screen out, grind up, or separate debris is the first step in wastewater treatment. Sticks, rags, large food particles, sand, gravel, toys, etc., are removed at this stage to protect the pumping and other equipment in the treatment plant. Treatment equipment such as bar screens, comminutors (a large version of a garbage disposal), and grit chambers are used as the wastewater first enters a treatment plant. The collected debris is usually disposed of in a landfill.

Primary Treatment

Primary treatment is the second step in treatment and separates suspended solids and greases from wastewater. Waste-water is held in a quiet tank for several hours allowing the particles to settle to the bottom and the greases to float to the top.

The solids drawn off the bottom and skimmed off the top receive further treatment as sludge. The clarified wastewater flows on to the next stage of wastewater treatment. Clarifiers and septic tanks are usually used to provide primary treatment.

Secondary Treatment

Secondary treatment is a biological treatment process to remove dissolved organic matter from wastewater. Sewage microorganisms are cultivated and added to the wastewater. The microorganisms absorb organic matter from sewage as their food supply. Three approaches are used to accomplish secondary treatment; fixed film, suspended film and lagoon systems.

Fixed Film Systems

Fixed film systems grow microorganisms on substrates such as rocks, sand or plastic. The wastewater is spread over the substrate, allowing the wastewater to flow past the film of microorganisms fixed to the substrate.

As organic matter and nutrients are absorbed from the wastewater, the film of microorganisms grows and thickens. Trickling filters, rotating biological contactors, and sand filters are examples of fixed film systems.

Suspended Film Systems

Suspended film systems stir and suspend microorganisms in wastewater. As the microorganisms absorb organic matter and nutrients from the wastewater they grow in size and amount. After the microorganisms have been suspended in the wastewater for several hours, they are settled out as a sludge.

Some of the sludge is pumped back into the incoming wastewater to provide "seed" microorganisms. The remainder is wasted and sent on to a sludge treatment process. Activated sludge, extended aeration, oxidation ditch, and sequential batch reactor systems are all examples of suspended film systems.

Lagoon Systems

Lagoon systems are shallow basins which hold the waste-water for several months to allow for the natural degradation of sewage. These systems take advantage of natural aeration and microorganisms in the wastewater to renovate sewage.

Final Treatment

Final treatment focuses on removal of disease-causing organisms from wastewater. Treated wastewater can be disinfected by adding chlorine or by using ultraviolet light. High levels of chlorine may be harmful to aquatic life in receiving streams. Treatment systems often add a chlorine-neutralizing chemical to the treated wastewater before stream discharge.

Sludge Treatment Advanced Treatment

Advanced treatment is necessary in some treatment systems to remove nutrients from wastewater. Chemicals are sometimes added during the treatment process to help settle out or strip out phosphorus or nitrogen. Some examples of nutrient removal systems include coagulant addition for phosphorus removal and air stripping for ammonia removal.

Sludges

Sludges are generated through the sewage treatment process. Primary sludges, material that settles out during primary treatment, often have a strong odour and require treatment prior to disposal. Secondary sludges are the extra microorganisms from the biological treatment processes.

The goals of sludge treatment are to stabilize the sludge and reduce odors, remove some of the water and reduce volume, decompose some of the organic matter and reduce volume, kill disease causing organisms and disinfect the sludge. Untreated sludges are about 97 per cent water. Settling the sludge and decanting off the separated liquid removes some of the water and reduces the sludge volume. Settling can result in a sludge with about 96 to 92 per cent water. More water can be removed from sludge by using sand drying beds, vacuum filters, filter presses, and centrifuges resulting in sludges with between 80 to 50 per cent water.

This dried sludge is called a sludge cake. Aerobic and anaerobic digestion are used to decompose organic matter to reduce volume. Digestion also stabilizes the sludge to reduce odors. Caustic chemicals can be added to sludge or it may be heat treated to kill disease-causing organisms. Following treatment, liquid and cake sludges are usually spread on fields, returning organic matter and nutrients to the soil. Wastewater treatment processes require careful management to ensure the protection of the water body that receives the discharge. Trained and certified treatment plant operators measure and monitor the incoming sewage, the treatment process and the final effluent.

Wastewater Treatment Regulations

Clean water has been a concern nationwide Congress adopted the Clean Water Act to protect the waters of the nation. Through this act. Environmental Protection Agency and corresponding state agencies were given the responsibility to regulate activities that threaten the quality of the nation's water resources.

In the Federal Clean Water Act, Congress adopted a comprehensive water policy for the nation and set as a national goal the elimination of pollutant discharges to the navigable waters. An interim goal was set to insure that all navigable waters would be fishable and swimmable To reach these goals.

- No one has the right to pollute the navigable waters of the United States. Dischargers are required to obtain permits.
- Permits shall set limits on the concentration of the pollutants being discharged. A violation of the limits carries a penalty of fines or imprisonment.
- The best technology available shall be used to control the discharge of pollutants.

Each state, through a designated regulatory agency, issues discharge permits and enforces the discharge limits. The Ohio Water Quality Standards specify that all surface waters of the state must be free from the following pollutants as a result of human activity: suspended solids, floating debris, colour, odour, toxic substances and nutrients that create nuisance growths of weeds and algae. Permits for wastewater treatment systems serving single family homes, two- or three-family dwellings are issued by local health departments. Permits for all wastewater treatment systems serving more than 2 homes or buildings larger than a 3-family dwelling are issued by the Ohio Environmental Protection Agency. Two different permits are issued by the Ohio Environmental Protection Agency. All systems must first obtain a Permit to Install (PTI) before a treatment plant can be constructed.

The Permit to Install is issued after the plans for construction, operation and management are examined to insure that the receiving stream will be protected. If the discharge from the treatment plant enters a stream, a NPDES permit is required. The NPDES permit specifies the maximum allowable level of total suspended solids, biochemical oxygen demand, nutrients and bacteria that can be discharged to a stream as well as the minimum level of dissolved oxygen that must be present in the discharge.

The levels specified in the NPDES permit are determined by the condition of the receiving stream. Therefore, NPDES permits are subject to change every 5 years as water quality concerns change throughout Ohio. Discharging raw sewage from a dwelling is considered a public nuisance. For systems serving a 1-, 2- or 3-family dwelling, the local health department is the responsible regulatory agency. The penalty for creating a public nuisance is a third degree misdemeanor which is punishable by not more than 60 days in jail and/or a $500 fine. In addition, the violator may be required to remove the sewage from public or private property or the waters of the state.

Discharges of pollutants from a multiple dwelling is under the jurisdiction of the Ohio Environmental Protection Agency. Discharges may be liable for civil penalties of up to $10,000 for each day of violation. In addition, criminal penalties can be assessed or up to one year of imprisonment or both. A "Connection Ban" can also be issued by the Ohio Environmental Protection Agency which prohibits the construction or installation of home sewage disposal systems when it is shown that pollutants from new homes will be discharged into the waters of the state.

The Ohio Environmental Protection Agency issued 25 administrative orders totaling $80,000 in fines to Ohio communities for wastewater treatment violations.This publication may contain pesticide recommendations that are subject to change at any time. These recommendations are provided only as a guide. It is always the pesticide applicator's responsibility, by law, to read and follow all current label directions for the specific pesticide being used. Due to constantly changing labels and product registrations, some of the recommendations given in this writing may no longer be legal by the time you read them. If any information in these recommendations disagrees with the label, the recommendation must be disregarded.

COMPOSTING

Composting refers to the process of combining organic materials such as weeds, grass clippings, and food scraps under conditions that enhance the rate at which they decompose. In the United States, organic wastes make up 20 per cent to 40 per cent of what we throw away. Composting provides an opportunity to keep these organic materials out of landfills and incinerators, using them instead to create a useful end product. Of course, composting isn't as easy as simply throwing organic waste into big piles and waiting for it to decompose.

If moist food scraps or grass clippings are placed in a container and left to sit for a week or two, the result is likely to be a stinky mess that attracts flies. Given the proper conditions, these same organic materials can be composted to produce a material that looks and smells like rich soil and is prized by gardeners. Clearly the second option is preferable, and the key question is how to provide the proper conditions for problem-free composting.

The answer lies in paying attention to moisture content, airflow, and the mixture between materials that are high in carbon and those that are high in nitrogen.. Composting takes place at a wide range of scales, from huge commercial or municipal operations to simple backyard bins or heaps. It can also be carried out indoors in containers such as garbage can bioreactors or worm bins. For research purposes, composting can even take place in soda bottle bioreactors, which are small and inexpensive enough to enable students to build their own individualized systems. Once made, compost can be used for gardening projects or for plant growth experiments ranging from nutrient analysis of compost-enriched soils to use of composts to suppress plant diseases.

WASTE MANAGEMENT AND POLLUTION CONTROL

Pollution abatement plans will not specify the technology to be used for reducing the pollution, but rather the maximum quantitative limits that are permissible and the corresponding time periods. Limits may be imposed either by source (*e.g.*, firm), or, as an option for certain classes of pollutants, by environmental sink, such as rivers, aquifers and the soils in defined zones. When the limits correspond to an entire sink, the Government may consider instituting systems of tradeable pollution rights for given classes of pollutants. Direct controls on abatement technologies have been shown to be unduly costly ways to reduce pollution.

The collection and disposal of solid waste, which is limited to the Georgetown area, currently is free of charge, but it needs improvement. Households improperly dispose of solid wastes owing to lags in collection times and non-maintenance of collection schedules for residential customers. In addition, an estimated 37 per cent of garbage in Georgetown remains uncollected. The necessary improvement in solid waste management can be combined with introduction of collection fees or privatisation of the service through contracting it out.

This would be a critical step in providing environmental services at a level that prevents grave public health hazards. Solid waste management programmes will be drawn up by the EPA, made available for comment in draft, and, once approved, compliance with their regulations will be mandatory, as in the case of other classes of pollutants. These programmes will also specify measures and investments required to increase waste management capacities.

Whenever possible, day-to-day waste management at the city and neighbourhood level will be contracted out to private entities, while the public sector will maintain an overall supervisory role and will be responsible for seeing that adequate waste disposal sites are available. As with other public services, an inadequacy of operating budgets has also troubled the water and sewerage sector. Yet there is abundant scope for recovering costs in this sector through the application of the appropriate charges. The prevailing picture is one of low water tariffs and low collection rates across the board for all uses of water. The water rate is as low as US$0.60 for some consumers. Outside the municipalities, the Regional Water Departments (RWD) manage to collect only from those who are willing to pay. Since the proceeds are deposited in the general revenue account and the RWD's operations are funded out of block grants from the Central Government, RWD do not have an incentive for financial self sufficiency.

Therefore, they have no interest in undertaking proper billings or strengthening collecting performance. In 1992, cost recovery excluding depreciation, was as low as 3.4 per cent in Regions 2 to 7. Clearly financial autonomy must be considered for these institutions. In keeping with the priority given to public health issues, emphasis will be given to replacing over time Georgetown's reliance on groundwater with piped water from upstream supplies. Investment priorities will include better sewage treatment facilities and/or extending Georgetown's current sewage outtake to several kilometres offshore.

AMERINDIAN ISSUES

A number of the issues outlined in the sectoral and cross-sectoral objectives discussed above have a specific impact on Amerindian communities. The issues mainly relate to rights to the land, benefits from exploitation of natural resources on that land and in the sub-soil, and the participation of Amerindians in decisions that affect their present and future.

Land and Exploitation Rights

To date the Amerindian communities have not received compensation for the exploitation of natural resources on land to which they lay claim. The National Development Strategy proposes that a legal basis for such claims be identified, as is currently proposed for agricultural land for all Guyanese, as well as an administrative system for the payment of royalties to Amerindian communities by concession holders who exploit natural resources on the lands to which the Amerindians lay claim.

Mining

Mining, whether alluvial or hardrock mining, river dredging or land mining causes disturbances to the environment that adversely affect the living conditions of Amerindians. To reduce the chances of degradation of the environment and erosion of the traditional way of life of Amerindians, the National Development Strategy proposes the formulation and implementation of a long-term land use plan which prohibits mining in Amerindian areas unless measures to limit degradation and to implement mitigation measures are guaranteed.

Intellectual Property Rights

To protect against the loss of intellectual property rights by foreign or private sector companies and research institutions, the activities of such companies in areas of traditional Amerindian knowledge should be monitored. Government will review the possibility of adopting the model provisions proposed by the World Intellectual Property Rights Organisation (WIPO) for protection against illicit exploitation of indigenous intellectual property.

Participatory Natural Resources Management

Amerindians are insufficiently involved in the management, administration and conservation of natural resources. Policies and mechanisms need to be established, possibly with assistance from NGOs with relevant experience, to grant substantially greater autonomy and foster greater community involvement in natural resource and environmental management.

RECOMMENDED LEGISLATIVE CHANGES

Guyana is a party to several multilateral treaties. However, the country has not been aggressive in bringing its domestic law into line with its international obligations, except as for UNCLOS III. This deficiency is particularly notable with regard to CITES. Guyana is also not a party to several multilateral treaties which are relevant to its circumstances. Notably these include the Ramsar Convention on Wetlands of international significance, the London Guideline for the exchange of information on chemicals in international trade, the Cartagena Convention on the marine environment in the Wider Caribbean Region, and the Kingston Protocol on Specially Protected Areas and Wildlife (SPAW).

A review of the legislation reveals that the law is relatively up to date only in the area of the management of non-renewable natural resources. Work remains to be done on legislation and regulations for the protection of wildlife, the designation and management of protected areas, the preparation of plans for control of municipal and industrial pollution and their enforcement (particularly in the coastal zone), the handling of agrochemicals, and the requirement of environmental impact assessments (EIAs) for development projects.

LEGISLATIVE REFORM PROGRAMME

Steps have been taken to remedy some deficiencies in the environmental laws and the following new legislative measures are under preparation:

- The Guyana Biosphere Reserves Bill, authorising the establishment and management of biosphere reserves in Guyana;
- The Conservation and Wildlife Bill, providing for the establishment of wildlife sanctuaries and the protection of listed wild animals and birds;
- The Pesticides and Toxic Chemical Control Bill, providing regulation of the importation, sale, and use of pesticides and toxic chemicals; and
- The Environmental Protection Bill, which creates an Environmental Protection Agency with powers to ensure effective management of natural resources and establish a regulatory regime for pollution control.

This Agency will have the powers to require that an Environmental Impact Assessment be submitted and reviewed for new activities. (Because of its importance, this Bill is reviewed in detail below.)

Given our national economic development agenda, it is urgent to move forward rapidly with legislative reform.

The following are areas in which legislative changes are necessary:

- Consolidation and revision of existing legislation in the various sectors, incorporating new environmental protection provisions.
- Preparation of consequential amendments to related legislation.
- Preparation of new legislation to repeal and replace existing but outdated Wildlife and National Parks and Protected Areas legislation.

It must be stressed, however, that the policy issues identified in this sectoral strategy must be considered in conjunction with the proposal for the Draft Environmental Protection Bill and the creation of the EPA. The functions of the EPA, the mechanisms for the execution of its mandate and the provisions specified must be the guiding principles for legislative reform and will serve the purpose of reconciling deficiencies and overlaps.

ENVIRONMENTAL PROTECTION BILL

Given the basic accepted policy of the need for the establishment of an EPA the functions of this soon-to-be-created agency as outlined in the Draft Environmental Protection Bill are:

- To take such steps as are necessary for the effective management of the natural environment to ensure conservation, protection and sustainable use of its natural resources;
- To promote the participation of members of the public in the process of integrating environmental concerns in planning for development on a sustainable basis;
- To coordinate the environmental management activities of all persons, organisations and agencies;
- To establish, monitor and enforce environmental regulations;
- To prevent or control environmental pollution;
- To coordinate an integrated coastal zone management programme;
- To ensure that any developmental activity which may cause an adverse effect on the natural environment be assessed before such activity is commenced and that such adverse effect be taken into account in deciding whether or not such activity should be authorised;
- To establish and maintain a programme for the conservation of biological diversity and its sustainable use;
- To coordinate the establishment and maintenance of a national parks and protected area system and a wildlife protection management programme;
- To promote and encourage a better understanding and appreciation of the natural environment and its role in the social and economic development;
- To establish and coordinate institutional linkages locally, nationally, regionally and internationally;
- To play a coordinating role in the preparation and implementation of cross-sectoral programmes with environmental content;
- To advise the Minister on matters of general policy relating to the protection, conservation and care of the environment and the impact of development;

- To perform such other functions pertaining to the protection of the environment as may be assigned to it by the Minister by or under this Act or any other law.

In the exercise of its functions the Agency may:

- Formulate and submit to the Minister policy recommendations and plans in furtherance of its functions;
- Conduct surveys, investigations and monitoring of the causes, nature, extent and prevention of harm to the environment and of pollution and assist and cooperate with other persons or bodies carrying out similar surveys or investigations;
- Conduct, promote and coordinate research in relation to any aspect of pollution or prevention thereof;
- Conduct investigations and inspections to ensure compliance with this Act or the regulations and investigate complaints relating to breaches of this Act or the regulations made thereunder;
- Provide information and education to the public regarding the need for methods of protection of the environment, improvement of the environment where altered directly or indirectly by human activity and the benefits of sustainable use of natural resources;
- Conduct and coordinate compilation of resource inventories, surveys and ecological analyses to monitor and obtain information on the social and biophysical environment with special reference to environmentally sensitive areas and areas where development is already taking place or likely to take place;
- Cormulate standards and codes of practice to be observed for the improvement and maintenance of the quality of the environment and limits on the release of contaminants into the environment;
- Request, examine, review, evaluate and approve or reject environmental impact assessments and risk analyses and make suitable recommendations for the mitigation of adverse effects of any proposed activity on the environment;
- Conduct studies and make recommendations on standards relating to the improvement of the environment and the maintenance of a sound ecological system;
- Monitor and coordinate monitoring of trends in the use of natural resources and their impact on the environment;
- Establish and enforce administrative penalties;
- Advise the Minister on the content and applicability of environmental control instruments;
- Produce sectoral guidelines on what may constitute significant effects on the environment;

- Obtain expert or technical advice from any suitably qualified person on such terms and conditions as the Agency shall think fit.

In the exercise of its functions the Agency also shall:

- Compile with the assistance of internationally recognised environmental groups a list of approved persons who have the qualifications and experience to carry out environmental impact assessments;
- Produce physical accounts in accordance with modern accounting standards to record the natural capital of Guyana;
- Carry out surveys and obtain baseline information on the natural resources including ecosystems and micro-ecosystems, population counts, species identification, location and condition and make such surveys, studies and information available to members of the public at their request and upon payment of the cost of photocopies;
- Provide general information to the public on the state of the environment by regular reports produced at least annually;
- Maintain and make available to members of the public during normal working hours a register of all environmental impact assessments carried out, environmental authorisations granted and other information in accordance with section 36.

In performing these functions, the Agency shall make use of current principles of environmental management, namely:

- *The"polluter pays principle"*: the polluter should bear the cost of measures to reduce pollution decided upon by public authorities to ensure that the environment is in an acceptable state, and should compensate citizens for the harm they suffer from pollution;
- *The"precautionary principle"*: where there are threats of serious or irreversible damage, lack of full scientific certainty shall not be used as a reason for postponing measures to prevent environmental degradation;
- The"strict liability" legal principle: any person who contravenes this Act or regulations shall be liable to the penalties prescribed thereafter;
- *The"avoidance" principle:* it is preferable to avoid environmental damage as it can be impossible or more expensive to repair rather than prevent damage;
- *The"state of technology" principle*: measures protecting the environment are restricted by what is technologically feasible and as technology improves, the improved technology should be used to prevent and repair environmental damage.

The Draft Environmental Protection Bill also addresses the matter of deference to authority of Agency by stating "Without prejudice to the provisions of section 14, any person or authority under any other written law, vested with power in relation to the environment shall defer to the authority of the Agency and shall request an environmental authorisation from the Agency before approving or determining any matter in respect of which an environmental authorisation is required under this Act."

2

Petroleum Hydrocarbon Contaminants

PETROLEUM HYDROCARBON

Petroleum, oily, flammable liquid that occurs naturally in deposits, usually beneath the surface of the earth; it is also called crude oil. It consists principally of a mixture of hydrocarbons

HYDROCARBON

Hydrocarbon any organic compound composed solely of the elements hydrogen and carbon. The hydrocarbons differ both in the total number of carbon and hydrogen atoms in their molecules and in the proportion of hydrogen to carbon. The hydrocarbons can be divided into various homologous series. Each member of such a series shows a definite relationship in its structural formula to the members preceding and following it, and there is generally some regularity in changes in physical properties of successive members of a series.

The alkanes are a homologous series of saturated aliphatic hydrocarbons. lThe first and simplest member of this series is methane, CH_4; the series is sometimes called the methane series. Each successive member of a homologous series of hydrocarbons has one more carbon and two more hydrogen atoms in its molecule than the preceding member. The second alkane is ethane, C_2H_6, and the third is propane, C_3H_8. Alkanes have the general formula C_nH_{2n+2} (where n is an integer greater than or equal to 1). Generally, hydrocarbons of low molecular weight, *e.g.*, methane, ethane, and propane, are gases; those of intermediate molecular weight, *e.g.*, hexane, heptane, and octane, are liquids; and those of high molecular weight, *e.g.*, eicosane ($C_{20}H_{42}$) and polyethylene, are solids.

Paraffin is a mixture of high-molecular-weight alkanes; the alkanes are sometimes called the paraffin series. Other homologous series of hydrocarbons include the alkenes alkene (alkin), any of a group of aliphatic hydrocarbons whose molecules contain one or more carbon-carbon double bonds Many common natural substances, *e.g.*, natural gas, petroleum, and asphalt, are complex mixtures of hydrocarbons. The coal tar obtained from coal by coking is also a mixture of hydrocarbons. Natural gas, petroleum, and coal tar are important sources of many hydrocarbons. These complex mixtures can be refined into simpler mixtures or pure substances by fractional distillation. During the refining of petroleum, one kind of hydrocarbon is often converted to another, more useful kind by cracking. Useful hydrocarbon mixtures include cooking gas, gasoline, naphtha, benzine, kerosene, paraffin, and lubricating oils.

Many hydrocarbons are useful as fuels; they burn in air to form carbon dioxide and water. The hydrocarbons differ in chemical activity. The alkanes are unaffected by many common reagents, while the alkenes and alkynes are much more reactive, as a result of the presence of unsaturation (*i.e.*, a carbon-carbon double or triple bond) in their molecules. Many important compounds are derived from hydrocarbons, either by substitution or replacement by some other chemical group or element of one or more of the hydrogen atoms of the hydrocarbon molecule, or by the addition of some element or group to a double or triple bond (in an unsaturated hydrocarbon). Such derivatives include alcohols, aldehydes, ethers, carboxylic acids, and halocarbons.

Alkane

Alkane any of a group of aliphatic hydrocarbons whose molecules contain only single bonds chemical bond, mechanism whereby atoms combine to form molecules. There is a chemical bond between two atoms or groups of atoms when the forces acting between them are strong enough to lead to the formation of an aggregate with sufficient stability to be regarded as. Alkanes have the general chemical formula C_nH_{2n+2}. An alkane is said to have a continuous chain if each carbon atom in its molecule is joined to at most two other carbon atoms; it is said to have a branched chain if any of its carbon atoms is joined to more than two other carbon atoms.

The first four continuous-chain alkanes are methane, ethane, propane, butane Names of continuous-chain alkanes whose molecules contain more than five carbon atoms are formed from a root that indicates the number of carbon atoms and the suffix -ane to indicate that the compound is an alkane; *e.g.*, alkanes with 5, 6, 7, 8, 9, and 10 carbon atoms in their molecules are pentane, hexane, heptane, octane, nonane, and decane, respectively. The name of a branched-chain alkane is formed by adding prefixes to the name of the continuous-chain alkane from which it is considered to be derived; *e.g.*, 2-methylpropane (called also isobutane) is thought of as being derived by replacing one of the hydrogen atoms bonded to the second (2-) carbon atom of a propane molecule with a methyl (CH_3) group, forming $CH_3CH(CH_3)_2$. Chemically, the alkanes are relatively unreactive. They are obtained by fractional distillation from petroleum and are used extensively as fuels. The alkanes are sometimes referred to as the methane series (after the simplest alkane) or as paraffins.

Methane

Methane,CH_4, colourless, odorless, gaseous saturated hydrocarbon; the simplest alkane any of a group of aliphatic hydrocarbons whose molecules contain only single bonds. It is less dense than air, melts at "184°C;, and boils at "161.4°C;. It is combustible and can form explosive mixtures with air. Methane occurs naturally as the principal component of natural gas; it is formed by the decomposition of plant and animal matter.

When this decomposition occurs underwater in swamps and marshes, marsh gas is released. The firedamp of coal mines is chiefly methane. In the atmosphere methane is a greenhouse gas, helping to trap infrared radiation, electromagnetic radiation having a wavelength in the range from c.75 × 10^{-6} cm to c.100,000 × 10^{-6} cm (0.000075–0.1 cm). Methane can be prepared in the labouratory by heating sodium acetate with sodium hydroxide, by the reaction of aluminum carbide with water, by the direct combination of carbon and hydrogen, or by the destructive distillation of coal or wood.

As natural gas, methane is widely used for fuel. It is also used for carbonizing steel. It is unaffected by many common chemical reagents but reacts violently with chlorine or fluorine in the presence of light and is therefore important as a starting material for the synthesis of solvents, *e.g.*, methylene chloride, chloroform, and carbon tetrachloride, and of some of the Freon refrigerants.

- Natural gas, natural mixture of gaseous hydrocarbons The composition of natural gas varies in different localities. Its chief component, methane, ethane, propane, butane and other hydrocarbon compounds. Some of the hydrocarbons found in gasoline gasoline or petrol, light, volatile mixture of hydrocarbons for use in the internal-combustion engine and as an organic solvent, obtained primarily by fractional distillation and "cracking" of petroleum, but also obtained from natural gas, by also occur as vapours in natural gas; by liquefying these hydrocarbons, gasoline can be obtained. Although commonly associated with petroleum petroleum, oily, flammable liquid that occurs naturally in deposits, usually beneath the surface of the earth; it is also called crude oil. It consists principally of a mixture of hydrocarbons, with traces of various nitrogenous and sulfurous compounds.. deposits it also occurs separately in sand, sandstone, and limestone deposits. Some geologists theorize that natural gas is a byproduct of decaying vegetable matter in underground strata, while others think it may be primordial gases that rise up from the mantle. Because of its flammability and high calorific value, natural gas is used extensively as an illuminant and a fuel.Natural gas was known to the ancients but was considered by them to be a supernatural phenomenon because, noticed only when ignited, it appeared as a mysterious fire bursting from the ground. One of the earliest attempts to harness it for economic use occurred in the early 19th cent. in Fredonia, N.Y. Towards the latter part of the 19th cent., large industrial cities began to make use of natural gas, and extensive pipeline systems have been constructed to transport gas.

- Liquefied natural gas, or LNG, is natural gas that has been pressurized and cooled so as to liquefy it for convenience in shipping and storage. The boiling point of natural gas is extremely low, and only in the 1970s did cryogenic technology physics low-temperature physics, science concerned with the production and maintenance of temperatures much below normal, down to almost absolute zero, and with various phenomena that occur only at such temperatures.advance enough to make the production and transport of LNG commerically feasible. Some of the natural gas moved to and from the United States is carried as LNG in special tankers.
- Infrared radiation, electromagnetic radiation, energy radiated in the form of a wave as a result of the motion of electric charges. A moving charge gives rise to a magnetic field, and if the motion is changing (accelerated), then the magnetic field varies and in turn produces an having a wavelength in the range from c.75 $\times 10^{-6}$ cm to c.100,000 $\times 10^{-6}$ cm (0.000075–0.1 cm). Infrared rays thus occupy that part of the electromagnetic spectrum arrangement or display of light or other form of radiation separated according to wavelength, frequency, energy, or some other property. Beams of charged particles can be separated into a spectrum according to mass in a mass spectrometer with a frequency less than that of visible light visible electromagnetic radiation. Of the entire electromagnetic spectrum, the human eye is sensitive to only a tiny part, the part that is called light. The wavelengths of visible light range from about 350 or 400 nm to about 750 or 800 nm.and greater than that of most radio waves, although there is some overlap. The name infrared means "below the red," *i.e.*, beyond the red, or lower-frequency (longer wavelength), end of the visible spectrum. Infrared radiation is thermal, or heat, no mechanical energy in transit, associated with differences in temperature between a system and its surroundings or between parts of the same system.
- Measures of Heat: It was first discovered in 1800 by Sir William Herschel, who was attempting to determine the part of the visible spectrum with the minimum associated heat in connection with astronomical observations he was making. In 1847, A. H. L. Fizeau and J. B. L. Foucault showed that infrared radiation has the same properties as visible light, being reflected, refracted, and capable of forming an interference interferometer. When the wavelength of the

light is known, the interferometer indicates the thickness of the film by the interference patterns it forms. The reverse process, *i.e.*, the measurement of the length of an unknown light wave, can also be carried out by the interferometer.

- Pattern: Infrared radiation is typically produced by objects whose temperature measure of the relative warmth or coolness of an object. Temperature is measured by means of a thermometer or other instrument having a scale calibrated in units called degrees. The size of a degree depends on the particular temperature scale being used. There are many applications of infrared radiation. A number of these are analogous to similar uses of visible light. Thus, the spectrum of a substance in the infrared range can be used in chemical analysis much as the visible spectrum is used. Radiation at discrete wavelengths in the infrared range is characteristic of many molecules. The temperature of a distant object can also be determined by analysis of the infrared radiation from the object. Radiometers operating in the infrared range serve as the basis for many instruments, including heat-seeking devices in missiles and devices for spotting and photographing persons and objects in the dark or in fog. Medical uses of infrared radiation range from the simple heat lamp to the technique of thermal imaging, or thermography. A thermograph of a person can show areas of the body where the temperature is much higher or lower than normal, thus indicating some medical problem. Thermography has also been used in industry and other applications. Some lasers produce infrared radiation. A recent development has been the expansion of research in infrared astronomy study of celestial objects by means of the infrared radiation they emit, in the wavelength range from about 1 micrometer to about 1 millimeter. All objects, from trees and buildings on the earth to distant galaxies, emit infrared (IR) radiation. Infrared sensors are sent aloft in balloons, rockets, and satellites to study the infrared radiation reaching the earth from other parts of the solar system and beyond.electromagnetic radiation, energy in physics, the ability or capacity to do work or to produce change.

Forms of energy include heat, light, sound, electricity, and chemical energy.. radiated in the form of a wave, in physics, the transfer of energy by the regular vibration, or oscillatory motion, either of some material medium or by the variation in magnitude of the field vectors of an electromagnetic field as a result of the motion of electric charges. A moving charge gives rise to a magnetic field, and if the motion is changing (accelerated), then the magnetic field varies and in turn produces an electric field. These interacting electric and magnetic fields are at right angles to one another and also to the direction of propagation of the energy. Thus, an electromagnetic wave is a transverse wave. If the direction of the electric field is constant, the wave is said to be polarized orientation of the vibration pattern of light waves in a singular plane.

DEVELOPMENT OF PETROLEUM

Petroleum has been known throughout historical time. It was used in mortar, for coating walls and boat hulls, and as a fire weapon in defensive warfare. Native Americans used it in magic and medicine and in making paints. Pioneers bought it from the Native Americans for medicinal use and called it Seneca oil and Genesee oil. In Europe it was scooped from streams or holes in the ground, and in the early 19th cent. small quantities were made from shale.

In 1815 several streets in Prague were lighted with petroleum lamps. Many wells were drilled in the region. Kerosene was the chief finished product, and kerosene lamps soon replaced whale oil lamps and candles in general use. Little use other than as lamp fuel was made of petroleum until the development of the gasoline engine and its application to automobiles, trucks, tractors, and airplanes.

Today the world is heavily dependent on petroleum for motive power, lubrication, fuel, dyes, drugs, and many synthetics. The widespread use of petroleum has created serious environmental problems. The great quantities that are burned as fuels generate most of the air pollution, contamination of the environment as a result of human activities. The term pollution refers primarily to the fouling of air, water, and land by wastes oil industry, the business of discovering oil (petroleum oily, flammable liquid that occurs naturally in deposits, usually beneath the surface of the earth; it is also called crude oil.

It consists principally of a mixture of hydrocarbons, with traces of various nitrogenous and sulfurous compounds.extracting it from the ground, refining it into a variety of products, and distributing it to the public. The development of the oil industry in the 19th and 20th cent. provided a source of energy that now supplies about two fifths of the world's energy needs as well as a raw material that chemical and petroleum industries refine into a number of essential chemicals and industrial products.

Petroleum seeping out of underground reservoirs has been collected and used for light throughout recorded history. In the 4th cent. A.D. the Chinese drilled for oil and natural gas, but in the 1850s, oil was still being recovered by skimming it off the tops of ponds. As whale oil became less abundant, producers looked for new ways to extract oil. Edwin Drake dug the first modern oil well in Titusville, Pa, hitting oil at 69.5 ft (21.2 m), touching off an oil rush in the area. (Most modern wells go down over 4,700 ft (1,432 m).) In 1861 the first oil refinery was set up.

Late-Twentieth Century and Early-Twenty-First Century Developments

OPEC required that the major oil companies provide them with a larger percentage of the profits from their fields. After the oil embargo in early1970's.The resulting energy crisis forced many developing countries to pay more for energy, negatively affecting Third World debt; industrialized countries implemented new measures to conserve and develop new sources of energy. With an abundant supply, oil prices dropped and stayed low through the 1990s, until early1990's when OPEC announced that it would cut production in order to increase oil prices worldwide. With the help of non-OPEC oil-producing nations, and world events and demand have driven prices significantly higher.

Economies dependent on oil production remain subject to the gyrations of the market. The collapse of oil prices in the mid-1980s ruined many independent refiners and helped produce a recession in such states as Texas. In contrast, the rise in oil prices early1990's has been responsible for economic growth in Russia, Venezuela, and other oil producers. Improved recovery methods combined with higher prices that justify more expensive extraction costs have rejuvenated production in some older oil fields, increased the estimates of reserves in existing fields, and made feasible the exploitation of deposits once considered uneconomical.Many oil-producing nations in the Middle East and Latin America have set up their own refining operations since the 1970s, and state-owned oil companies in OPEC countries are now among the world's largest.

Many large oil companies have diversified into chemicals, and oil prices are increasingly set on commodity trading exchanges such as the New York Mercantile Exchange. Beginning in the late 1990s, the industry saw increased consolidation as already large oil companies merged with each other, including Exxon with Mobil (the second largest; forming ExxonMobil), Chevron with Texaco and Unocal as Chevron, British Petroleum with Amoco and ARCO as BP, and Conoco with Phillips Petroleum as ConocoPhillips.

THE BOILING POINT AND LATENT HEAT OF VAPOURIZATION

When heat is added to a liquid at its boiling point boiling point, temperature at which a substance changes its state from liquid to gas. A stricter definition of boiling point is the temperature at which the liquid and vapour (gas) phases of a substance can exist in equilibrium., with the pressure kept constant, the molecules of the liquid acquire enough energy to overcome the intermolecular forces intermolecular forces, forces that are exerted by molecules on each other and that, in general, affect the macroscopic properties of the material of which the molecules are a part. Such forces may be either attractive or repulsive in nature. that bind them together in the liquid state, and they escape as individual molecules of vapour until the vapourization is complete. Vapourization at the boiling point is known simply as boiling. The temperature of a boiling liquid remains constant until all of the liquid has been converted to a gas. For each substance a certain specific amount of heat must be supplied to vapourize a given quantity of the substance. This amount of heat is known as the latent heat latent heat, heat change associated with a change of state or phase Latent heat, also called heat of transformation, is the heat given up or absorbed by a unit mass of a substance as it changes from a solid to a liquid, from a liquid to a of vapourization of the substance. The quantity of heat applied for each gram (or each molecule) undergoing the change in state depends on the substance itself.

The amount of heat necessary to change one gram of water to steam at its boiling point at one atmosphere of pressure, *i.e.*, the heat of vapourization of water, is approximately 540 calories. Other substances require other amounts. Boiling point, temperature at which a substance changes its state from liquid to gas. A stricter definition of boiling point is the temperature at which the liquid and vapour (gas) phases of a substance can exist in equilibrium.

When heat is applied to a liquid, the temperature of the liquid rises until the vapour pressure vapour pressure, pressure exerted by a vapour that is in equilibrium with its liquid. A liquid standing in a sealed beaker is actually a dynamic system: some molecules of the liquid are evapourating to form vapour and some molecules of vapour are condensing to form liquid.. of the liquid equals the pressure of the surrounding gases. At this point there is no further rise in temperature, and the additional heat energy supplied is absorbed as latent heat latent heat, heat change associated with a change of state or phase Latent heat, also called heat of transformation, is the heat given up or absorbed by a unit mass of a substance as it changes from a solid to a liquid, from a liquid to a. vapourization to transform the liquid into gas. This transformation occurs not only at the surface of the liquid (as in the case of evapouration change of a liquid into vapour at any temperature below its boiling point.

For example, water, when placed in a shallow open container exposed to air, gradually disappears, evapourating at a rate that depends on the amount of surface exposed, the humidity but also throughout the volume of the liquid, where bubbles of gas are formed. The boiling point of a liquid is lowered if the pressure of the surrounding gases is decreased. Water will boil at a lower temperature at the top of a mountain, where the atmospheric pressure on the water is less, than it will at sea level, where the pressure is greater. In the labouratory, liquids can be made to boil at temperatures far below their normal boiling points by heating them in vacuum flasks under greatly reduced pressure. On the other hand, if the pressure is increased, the boiling point is raised. For this reason, it is customary when the boiling point of a substance is given to include the pressure at which it is observed, if that pressure is other than standard, *i.e.*, 760 mm of mercury or 1 atmosphere. The boiling point of a solution solution, in chemistry, homogeneous mixture of two or more substances. The dissolving medium is called the solvent, and the dissolved material is called the solute.

A solution is distinct from a colloid or a suspension is always higher than that of the pure solvent; this boiling-point elevation is one of the colligative properties colligative properties, properties of a solution that depend on the number of solute particles present but not on the chemical properties of the solute. Colligative properties of a solution include freezing point.

EVAPOURATION AND VAPOUR PRESSURE

Liquids can also change to gases at temperatures below their boiling points. Vapourization of a liquid below its boiling point is called evapouration evapouration, change of a liquid into vapour at any temperature below its boiling point. Water, when placed in a shallow open container exposed to air, gradually disappears, evapourating at a rate that depends on the amount of surface exposed, the humidity. Which occurs at any temperature when the surface of a liquid is exposed in an unconfined space.

When, however, the surface is exposed in a confined space and the liquid is in excess of that needed to saturate the space with vapour, an equilibrium is quickly reached between the number of molecules of the substance going off from the surface and those returning to it. A change in temperature upsets this equilibrium; a rise in temperature, increases the activity of the molecules at the surface and consequently increases the rate at which they fly off.

When the temperature is maintained at the new point for a short time, a new equilibrium is soon established. The pressure exerted by the vapour of a liquid in a confined space is called its vapour pressure. It differs for different substances at any given temperature, but each substance has a specific vapour pressure for each given temperature. At its boiling point the vapour pressure of a liquid is equal to atmospheric pressure. The vapour pressure of water, measured in terms of the height of mercury in a barometer, is 4.58 mm at 0°C; and 760 mm at 100°C; (its boiling point).

THE BASIC DISTILLATION PROCESS

A simple distillation apparatus consists essentially of three parts: a flask equipped with a thermometer and with an outlet tube from which the vapour is emitted; a condenser that consists of two tubes of different diameters placed one within the other and so arranged that the smaller (in which the vapour is condensed) is held in a stream of coolant in the larger; and a vessel in which the condensed vapour is collected. The mixture of substances is placed in the flask and heated. Ideally, the substance with the lowest boiling point boiling point, temperature at which a substance changes its state from liquid to gas. A stricter definition of boiling point is the temperature at which the liquid and vapour (gas) phases of a substance can exist in equilibrium.vapourization, change of a liquid or solid substance to a gas or vapour. There is fundamentally no difference between the terms gas and vapour, but gas is used commonly to describe a substance that appears in the gaseous state under standard conditions ofthe temperature remaining constant until that substance has completely distilled. The vapour is led into the condenser where, on being cooled, it reverts to the liquid (condenses) and runs off into a receiving vessel. The product so obtained is known as the distillate.

Those substances having a higher boiling point remain in the flask and constitute the residue.Since a perfect separation is never effected, the distillate is often redistilled to increase its purity (hence the expression "double distilled" or "triple distilled"). Many alcoholic beverages are distilled, *e.g.*, brandy, gin, whiskey, and various liqueurs. The apparatus used, called the still, is the same in principle as other distillation apparatus.

THE FRACTIONAL DISTILLATION PROCESS

When the substance with the lowest boiling point has been removed, the temperature can be raised and the distillation process repeated with the substance having the next lowest boiling point. The process of obtaining portions (or fractions) in this way is one type of fractional distillation. A more efficient method of fractional distillation involves placing a vertical tube called a fractionating column between the flask and the condenser. The column is filled with many objects on which the vapour can repeatedly condense and reevapourate as it moves towards the top, effectively distilling the vapour many times. The less volatile substances in the vapour tend to run back down the column after they condense, concentrating themselves near the bottom. The more volatile ones tend to reevapourate and keep moving upward, concentrating themselves near the top.

Because of this the column can be tapped at various levels to draw off different fractions. Fractional distillation is commonly used in refining petroleum, some of the fractions thus obtained being gasoline, benzene, kerosene, fuel oils, lubricating oils, and paraffin.

THE DESTRUCTIVE DISTILLATION PROCESS

Another form of distillation involves heating out of free contact with air such substances as wood, coal, and oil shale and collecting separately the portions driven off; this is known as destructive distillation. Wood, when treated in this way yields acetic acid, methyl or wood alcohol, charcoal, and a number of hydrocarbons. Coal yields coal gas, coal tar, ammonia, and coke.

Ammonia is also obtained by the destructive distillation of oil shale. Petrochemical, any one of a large group of chemicals derived from a component of petroleum or natural gas. The cracking processes for manufacturing gasoline produce vast quantities of gaseous hydrocarbons.

Originally considered waste products suitable only for use as illuminants and fuels, the gases today are manufactured into petrochemical substances widely employed in industry. Important petrochemical compounds are alcohols and aldehydes, butylene, butadiene, ethylene, propylene, toluene, styrene, acetylene, benzene, ethylene oxide, ethylene glycol, acrylonitrile, acetone, acetic acid, acetic anhydride, and ammonia. Materials made from the gases include carbon black, synthetic rubber, polystyrene, polypropylene, and polyethylene. Petrochemicals are widely used in agriculture, in the manufacture of plastics, synthetic fibres, and explosives, and in the aircraft and automobile industries.

PHYSICAL PROPERTIES AND EXACT CHEMICAL COMPOSITION AND REFINING OF PETROLEUM

The physical properties and exact chemical composition of crude oil varies from one locality to another. The different hydrocarbon components of petroleum are dissolved natural gas Liquefied natural gas, or LNG, is natural gas that has been pressurized and cooled so as to liquefy it for convenience in shipping and storage. The boiling point of natural gas is extremely low, and only in the 1970s did cryogenic technology advance, gasoline, benzine, colourless, highly flammable liquid. Naphtha, kerosene or kerosine, colourless, thin mineral oil whose density is between 0.75 and 0.85 grams per cubic centimeter.

A mixture of hydrocarbons, it is commonly obtained in the fractional distillation of petroleum as the portion boiling off., diesel fuel and light heating oils, heavy heating oils, and finally tars of various weights tar and pitch, viscous, dark-brown to black substances obtained by the destructive distillation of coal, wood, petroleum, peat, and certain other organic materials.. The crude oil is usually sent from a well to a refinery in pipelines, hollow structure, usually cylindrical, for conducting materials. It is used primarily to convey liquids, gases, or solids suspended in a liquid, *e.g.*, a slurry. It is also used as a conduit for electric wires.

Or tanker ships.The hydrocarbon components are separated from each other by various refining processes. In a process called fractional distillation distillation, process used to separate the substances composing a mixture. It involves a change of state, as of liquid to gas, and subsequent condensation. The process was probably first used in the production of intoxicating beverages.. petroleum is heated and sent into a tower.

The vapours of the different components condense on collectors at different heights in the tower. The separated fractions are then drawn from the collectors and further processed into various petroleum products. One of the many products of crude oil is a light substance with little colour that is rich in gasoline or petrol, light, volatile mixture of hydrocarbons for use in the internal-combustion engine and as an organic solvent, obtained primarily by fractional distillation and "cracking" of petroleum, but also obtained from natural gas, by. Another is a black tarry substance that is rich in asphalt brownish-black substance used commonly in road making, roofing, and waterproofing.As the lighter fractions, especially gasoline, are in the greatest demand, so-called cracking processes have been developed in which heat, pressure, and certain catalysts are used to break up the large molecules of heavy hydrocarbons into small molecules of light hydrocarbons.

Some of the heavier fractions find eventual use as lubricating oils, paraffins white, more-or-less translucent, odorless, tasteless, waxy solid. It melts between 47°C; and 65°C; and is insoluble in water but soluble in ether, benzene, and certain esters., and highly refined medicinal substances such as petrolatum..

Benzine

Benzine colourless, highly flammable liquid. It is used as a cleaning agent because it is a solvent for organic substances such as fats, oils, and resins and is also used in the preparation of certain dyes and paints. Benzine is a mixture of hydrocarbons, chiefly alkanes such as pentane and hexane. It is obtained by the fractional distillation of petroleum oily, flammable liquid that occurs naturally in deposits, usually beneath the surface of the earth; it is also called crude oil. It consists principally of a mixture of hydrocarbons, with traces of various nitrogenous and sulfurous compounds.

Naphtha

Naphtha, term usually restricted to a class of colourless, volatile, flammable liquid hydrocarbon mixtures. Obtained as one of the more volatile fractions in the fractional distillation of petroleum petroleum, oily, flammable liquid that occurs naturally in deposits, usually beneath the surface of the earth; it is also called crude oil. It consists principally of a mixture of hydrocarbons, with traces of various nitrogenous and sulfurous compounds. (when it is known as petroleum naphtha), in the fractional distillation of coal tar (coal-tar naphtha), and in a similar distillation of wood (wood naphtha), it is used widely as a solvent for various organic substances, such as fats and rubber, and in the making of varnish. Because of its dissolving property it is important as a cleaning fluid; it is also incorporated in certain laundry soaps. Coal-tar (aromatic) naphthas have greater solvent power than petroleum (aliphatic) naphthas. Originally the term naphtha designated a colourless flammable liquid obtained from the ground in Persia. Later it came to be applied to a number of other natural liquid substances having similar properties. Technically, gasoline and kerosene are considered naphthas.

Kerosene or Kerosine

Kerosene or kerosine, colourless, thin mineral oil whose density is between 0.75 and 0.85 grams per cubic centimeter. A mixture of hydrocarbons, it is commonly obtained in the fractional distillation of petroleum as the portion boiling off between 150°C; and 275°C; (302°F; – 527°F;). Kerosene has been recovered from other substances, notably coal (hence another name, coal oil), oil shale, and wood. At one time kerosene was the most important refinery product because of its use in lamps. Now it is most noted for its use as a carrier in insecticide sprays and as a fuel in jet engines.

Paraffin

Paraffin, white, more-or-less translucent, odorless, tasteless, waxy solid. It melts between 47°C; and 65°C; and is insoluble in water but soluble in ether, benzene, and certain esters. Paraffin is unaffected by most common chemical reagents but burns readily in air. Obtained from petroleum during refining, it is used in candles, for coating paper, and for various other purposes. Chemically, paraffin is a mixture of high-molecular-weight alkanes, any of a group of aliphatic hydrocarbons whose molecules contain only single bonds *i.e.,* saturated hydrocarbons any organic compound composed solely of the elements hydrogen and carbon. With the general formula CnH2n+2, where n is an integer between 22 and 27.

Petrolatum

Petrolatum colourless to yellowish-white hydrocarbon mixture obtained by fractional distillation of petroleum. In its jellylike semisolid form (known as petroleum jelly and also by several trade names) it is used in preparing medicinal ointments and for lubrication. As a nearly colourless, highly refined liquid known as liquid petrolatum, liquid paraffin, or mineral oil, it is used as a lubricant, as a laxative, and as a base for nasal sprays.

Asphalt

Asphalt brownish-black substance used commonly in road making, roofing, and waterproofing. Chemically, it is a natural mixture of hydrocarbons. It varies in consistency from a solid to a semisolid, has great tenacity, melts when heated, and when ignited will burn with a smoky flame leaving very little or no ash. It is found in nature in deposits called asphalt lakes. Natural asphalt was probably formed by the evapouration of petroleum. Asphalt is obtained as a residue in the distillation or refining of petroleum. This is its important commercial source. It occurs also in asphalt rock, a natural mixture of asphalt with sand and limestone, which when crushed is used as road-building material. Asphalt is also used in the manufacture of paints and varnishes, giving an intensely black colour.

Gasoline or Petrol

Gasoline or petrol, light, volatile mixture of hydrocarbons for use in the internal-combustion engine internal-combustion engine, one in which combustion of the fuel takes place in a confined space, producing expanding gases that are used directly to provide mechanical power. and as an organic solvent, obtained primarily by fractional distillation and "cracking" of petroleum, but also obtained from natural gas, by destructive distillation of oil shales and coal, and by a process that converts methanol to gasoline using zeolite as a catalyst. Gasoline intended for use in engines is rated by octane number octane number, figure of merit representing the resistance of gasoline to premature detonation when exposed to heat and pressure in the combustion chamber of an internal-combustion engine., an index of quality that reflects the ability of the fuel to resist detonation and burn evenly when subjected to high pressures and temperatures inside an engine. Premature detonation produces "knocking" and "pinging"; it wastes fuel and may cause engine damage.

The addition of tetramethyl lead and tetraethyl lead to raise the octane number is no longer permitted in the United States because it leads to dangerous emissions containing lead. New formulations of gasoline designed to raise the octane number contain increasing amounts of aromatics and oxygen-containing compounds (oxygenates), such as alcohols, methyl tertiary butyl ether (MTBE), and methylcyclopentadienyl manganese tricarbonyl (MMT). Automobiles automobile, self-propelled vehicle used for travel on land.

The term is commonly applied to a four-wheeled vehicle designed to carry two to six passengers and a limited amount of cargo, as contrasted with a truck, which is designed primarily for the transportation are now equipped with catalytic converters that oxidize unreacted gasoline; the cars are designed to run on newly formulated gasolines as well as on gasohol gasohol, a gasoline extender made from a mixture of gasoline (90 per cent) and ethanol (10 per cent; often obtained by fermenting agricultural crops or crop wastes) or gasoline (97 per cent) and methanol, or wood alcohol (3 per cent)., which contains 10 per cent ethanol or 3 per cent methanol.

In addition, since 1998 a number of American automobiles have been equipped to enable them to run on either gasoline or E85, a mixture of 85 per cent ethanol and 15 per cent gasoline. Some racing cars use pure methanol methanol, methyl alcohol, or wood alcohol, CH_3OH, a colourless, flammable liquid that is miscible with water in all proportions. Methanol is a monohydric alcohol There are five blends of gasoline marketed in the United States.

Conventional gasoline, the most widely available, is sold where air quality is satisfactory; it has been formulated to evapourate more slowly in hot weather so as to reduce smog, and it now contains detergent additives to reduce engine deposits. Winter oxygenated gasoline, is formulated as conventional gasoline with oxygen-rich chemicals added, such as MTBE or ethanol. The oxygen promotes cleaner burning, reducing carbon monoxide, and is generally sold from November to March because cold engines operate less efficiently and produce more carbon monoxide. Reformulated gasoline

(RFG), is mandated in areas where toxins in the air are a constant problem; it contains oxygen-rich chemicals in lesser concentrations than the winter oxygenated gasoline and is formulated to reduce certain toxic chemicals found in conventional and winter oxygenated fuels.

Oxygenated reformulated gasoline is a wintertime fuel exclusive to the New York City area, where heavy carbon monoxide pollution occurs. California reformulated gasoline, introduced has a different formulation and burns cleaner than regular reformulated gasoline. Because MTBE has been implicated as a pollutant, particularly of groundwater, its use is being curtailed. California ruled that the MTBE in California reformulated gas must be phased.

Distillation

Distillation, process used to separate the substances composing a mixture. It involves a change of state, as of liquid to gas, and subsequent condensation condensation, in physics, change of a substance from the gaseous (vapour) to the liquid state Condensation is the reverse of vapourization, or change from liquid to gas.. The process was probably first used in the production of intoxicating beverages. Today, refined methods of distillation are used in many industries, including the alcohol and petroleum industries.

Condensation

Condensation, in physics, change of a substance from the gaseous (vapour) to the liquid state, forms of matter differing in several properties because of differences in the motions and forces of the molecules of which they are composed.

Vapourization

Vapourization, change of a liquid or solid substance to a gas or vapour. There is fundamentally no difference between the terms gas and vapour, but gas is used commonly to describe a substance that appears in the gaseous state under standard conditions of or change from liquid to gas. It can be brought about by cooling, as in distillation, process used to separate the substances composing a mixture. It involves a change of state, as of liquid to gas, and subsequent condensation.

The process was probably first used in the production of intoxicating beverages.or by an increase in pressure resulting in a decrease in volume. Certain natural phenomena, such as dew, fog, mist, and clouds, are the result of the condensation of water vapour in the atmosphere; the formation of dew, thin film of water that has condensed on the surface of objects near the ground. Dew forms when radiational cooling of these objects during the nighttime hours also cools the shallow layer of overlying air in contact with them, causing the condensation of some emphsises well the fundamental principles involved in such phenomena. The explanation of condensation can be found in the kinetic-molecular theory of gases, physical theory that explains the behaviour of gases on the basis of the following assumptions:

- Any gas is composed of a very large number of very tiny particles called molecules;
- The molecules are very far apart compared to. As heat is removed from a gas, the molecules of the gas move more slowly The intermolecular forces that are exerted by molecules on each other and that, in general, affect the macroscopic properties of the material of which the molecules are a part. Such forces may be either attractive or repulsive in nature. Are strong enough to pull the molecules together to form droplets of liquid. Similarly, reducing the volume of the gas reduces the average distance between molecules and thus favours the intermolecular forces tending to pull them together. Vapourization, change of a liquid or solid substance to a gas or vapour. There is fundamentally no difference between the terms gas and vapour, but gas is used commonly to describe a substance that appears in the gaseous state under standard conditions of pressure and temperature, and vapour to describe the gaseous state of a substance that appears ordinarily as a liquid or solid. Although most substances undergo changes of state in the order of solid to liquid to gas as the temperature is raised, a few change directly from solid to gas in a process known as sublimation sublimation, change of a solid substance directly to a vapour without first passing through the liquid state.

ORIGIN AND NATURAL OCCURRENCE

During the past 600 million years incompletely decayed plant and animal remains have become buried under thick layers of rock. It is believed that petroleum consists of the remains of these organisms but it is the small microscopic plankton organism remains that are largely responsible for the relatively high organic carbon content of fine-grained sediments like the Chattanooga shale which are the principle source rocks for petroleum.

Among the leading producers of petroleum are Saudi Arabia, Russia, the United States (chiefly Texas, California, Louisiana, Alaska, Oklahoma, and Kansas), Iran, China, Norway, Mexico, Venezuela, Iraq, Great Britain, the United Arab Emirates, Nigeria, and Kuwait. The largest known reserves are in the Middle East.

EXPLORATION AND DRILLING OF WELLS

Because of the subterranean origin of petroleum it must be extracted by means of wells well, aperture in the earth's surface through which substances in a natural underground reservoir, such as water, gas, oil, salt, and sulfur, can flow or be pumped to the surface. Until an exploratory well, or wildcat, has been dug, there is no sure way of knowing whether or not petroleum lies under a particular site. In order to reduce the number of exploratory wells drilled, scientific methods are used to pick the most promising sites. Sensitive instruments, such as the gravimeter, the magnetometer, and the seismograph, may be used to find subsurface rock formations that can hold crude oil. Drilling is a fairly complex and often risky process. Some wells must be dug several miles deep before petroleum deposits are reached. Many are now drilled offshore from platforms standing in the ocean bed.

Usually the petroleum from a new well will come to the surface under its own pressure. Later the crude oil must be pumped out or forced to the surface by injecting water, air, natural gas, steam, carbon dioxide, or another substance into the deposits. Enhanced recovery techniques have increased the percentage of oil that can be extracted from a field.

Well

Well, aperture in the earth's surface through which substances in a natural underground reservoir, such as water, gas, oil, salt, and sulfur, can flow or be pumped to the surface. In the United States, until some years after the Civil War, the majority of wells were "open," *i.e.*, holes dug in the ground and lined, or cased, with brick, stone, or wood. Although they are sometimes dug with picks and shovels, most wells today are made by rotary or percussion drills. An artesian well artesian well, deep drilled well through which water is forced upward under pressure. The water in an artesian well flows from an aquifer, which is a layer of very porous rock or sediment, usually sandstone, capable of holding and transmitting large quantities of the most desirable type of water well, is always drilled because rock layers must be cut through to reach the water. Oil wells are usually drilled using a rotary-drill method, in which a drilling bit set in the bottom of a drilling pipe is rotated by machinery on the ground level. As the cut deepens, more sections of pipe are fastened to the sections already in use. A special mixture called drilling mud is sent down through the pipe to wash away the drillings and also to cool the cutting bit. Some oil wells are drilled by a percussion method known as cable-tool drilling. In this procedure a heavy metal bit attached to a cable is alternately raised and dropped, pulverizing the rock beneath it. Water is pumped into the well and mixed with the rock cuttings, the mixture being bailed out when it becomes thick enough to interfere with the action of the bit.

Regardless of the drilling method, well walls are usually cased with iron or steel to prevent cave-ins. Casing is inserted when the desired depth has been reached or, in some instances, as the well is being drilled. Minerals, such as salt and sulfur, can be pumped to the surface through a well if they are first liquefied by some process; salt may be brought up if water is first pumped to the bottom of the well to dissolve the salt.

Artesian Well

Artesian well, deep drilled well well, aperture in the earth's surface through which substances in a natural underground reservoir, such as water, gas, oil, salt, and sulfur, can flow or be pumped to the surface through which water is forced upward under pressure. The water in an artesian well flows from an aquifer, which is a layer of very porous rock or sediment, usually sandstone, capable of holding and transmitting large quantities of water.

The geologic conditions necessary for an artesian well are an inclined aquifer sandwiched between impervious rock layers above and below that trap water in it. Water enters the exposed edge of the aquifer at a high elevation and percolates downward through interconnected pore spaces. The water held in these spaces is under pressure because of the weight of water in the portion of the aquifer above it. If a well is drilled from the land surface through the overlying impervious layer into the aquifer, this pressure will cause the water to rise in the well.

In areas where the slope of the aquifer is great enough, pressure will drive the water above ground level in a spectacular, permanent fountain. Artesian springs can occur in similar fashion where faults or cracks in the overlying impervious layer allow water to flow upward. Water from an artesian well or spring is usually cold and free of organic contaminants, making it desirable for drinking. In North America, the Dakota sandstone provides aquifers for an artesian system that underlies parts of the Dakotas, Montana, Wyoming, Kansas, Nebraska, and Saskatchewan and supplies great quantities of water to the dry Great Plains region.

Many East Coast cities derive their water supplies from aquifers that are exposed along the edge of the Piedmont and dip downward towards the Atlantic coast. The largest artesian system in the world underlies nearly all of E and S Australia. Other important artesian systems serve London, Paris, and E Algeria.

3

Treatment of Industrial Wastewater

The different types of contamination of wastewater require a variety of strategies to remove the contamination.

SOLIDS REMOVAL

Most solids can be removed using simple sedimentation techniques with the solids recovered as slurry or sludge. Very fine solids and solids with densities close to the density of water pose special problems. In such case filtration or ultra-filtration may be required. Alternatively, flocculation may be used using alum salts or the addition of poly-electrolytes

OILS AND GREASE REMOVAL

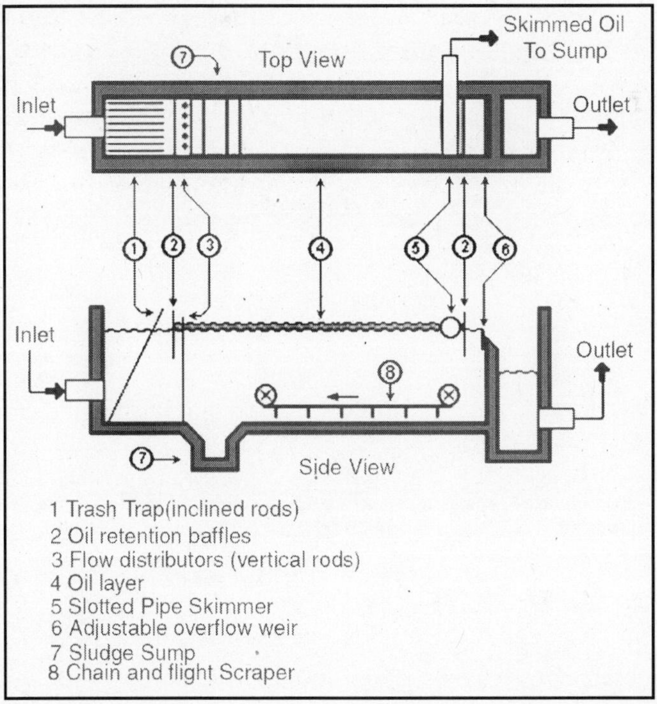

Fig. A Typical API Oil-water Separator used in Many Industries

Many oils can be recovered from open water surfaces by skimming devices. However, hydraulic oils and the majority of oils that have degraded to any extent will also have a soluble or emulsified component that will require further treatment to eliminate. Dissolving or emulsifying oil using surfactants or solvents usually exacerbates the problem rather than solving it, producing wastewater that is more difficult to treat. The wastewaters from large-scale industries such as oil refineries, petrochemical plants, chemical plants, and natural gas processing plants commonly contain gross amounts of oil and suspended solids.

Those industries use a device known as an API oil-water separator which is designed to separate the oil and suspended solids from their wastewater effluents. The name is derived from the fact that such separators are designed according to standards published by the American Petroleum Institute (API). The API separator is a gravity separation device designed by using Stokes Law to define the rise velocity of oil droplets based on their density and size.

The design is based on the specific gravity difference between the oil and the wastewater because that difference is much smaller than the specific gravity difference between the suspended solids and water. The suspended solids settles to the bottom of the separator as a sediment layer, the oil rises to top of the separator and the cleansed wastewater is the middle layer between the oil layer and the solids.

Typically, the oil layer is skimmed off and subsequently re-processed or disposed of, and the bottom sediment layer is removed by a chain and flight scraper (or similar device) and a sludge pump. The water layer is sent to further treatment consisting usually of a dissolved air flotation (DAF) unit for additional removal of any residual oil and then to some type of biological treatment unit for removal of undesirable dissolved chemical compounds.

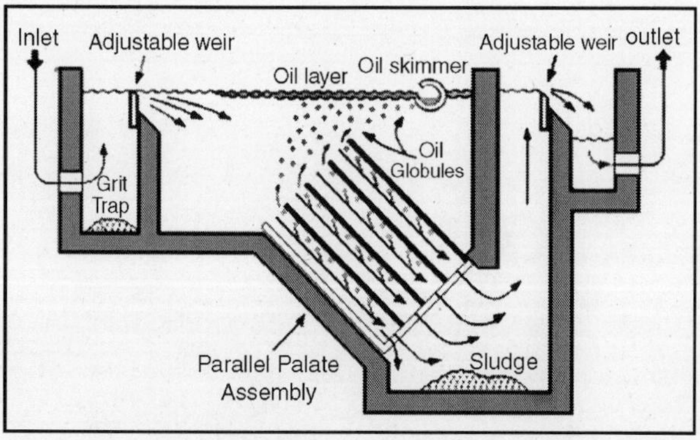

Fig. A Typical Parallel Plate Separator

Parallel plate separators are similar to API separators but they include tilted parallel plate assemblies (also known as parallel packs). The parallel plates provide more surface for suspended oil droplets to coalesce into larger globules. Such separators still depend upon the specific gravity between the suspended oil and the water.

However, the parallel plates enhance the degree of oil-water separation. The result is that a parallel plate separator requires significantly less space than a conventional API separator to achieve the same degree of separation.

REMOVAL OF BIODEGRADABLE ORGANICS

Biodegradable organic material of plant or animal origin is usually possible to treat using extended conventional wastewater treatment processes such as activated sludge or trickling filter.

Problems can arise if the wastewater is excessively diluted with washing water or is highly concentrated such as neat blood or milk. The presence of cleaning agents, disinfectants, pesticides, or antibiotics can have detrimental impacts on treatment processes.

Activated Sludge Process

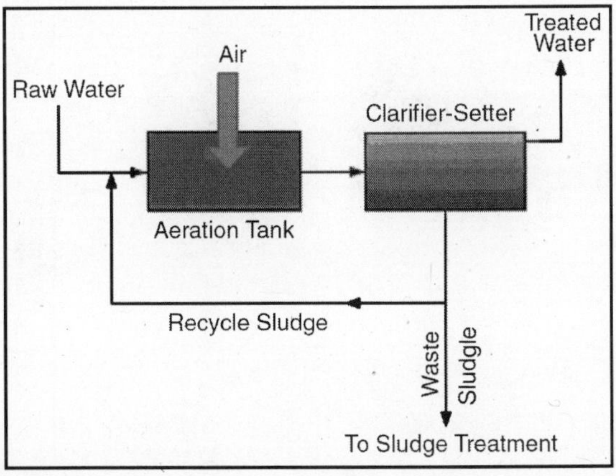

Fig. A Generalized, Schematic Diagram of an Activated Sludge Process.

Activated sludge is a biochemical process for treating sewage and industrial wastewater that uses air (or oxygen) and microorganisms to biologically oxidize organic pollutants, producing a waste sludge (or floc) containing the oxidized material. In general, an activated sludge process includes:

- An aeration tank where air (or oxygen) is injected and thoroughly mixed into the wastewater.
- A settling tank (usually referred to as a "clarifier" or "settler") to allow the waste sludge to settle. Part of the waste sludge is recycled to the aeration tank and the remaining waste sludge is removed for further treatment and ultimate disposal.

Trickling Filter Process

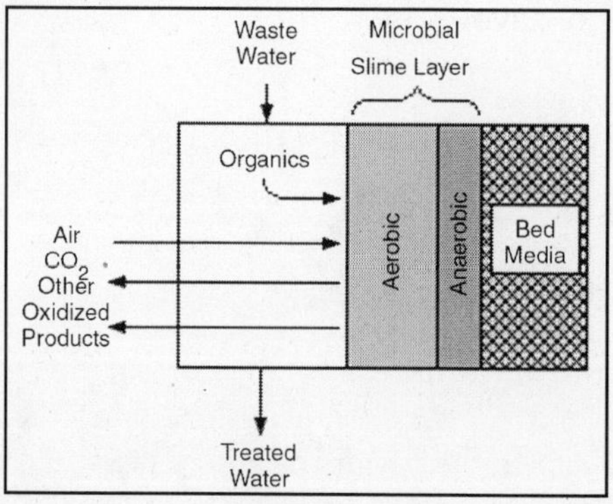

Fig. A Schematic Cross-section of the Contact face of the Bed Media in a Trickling Filter+

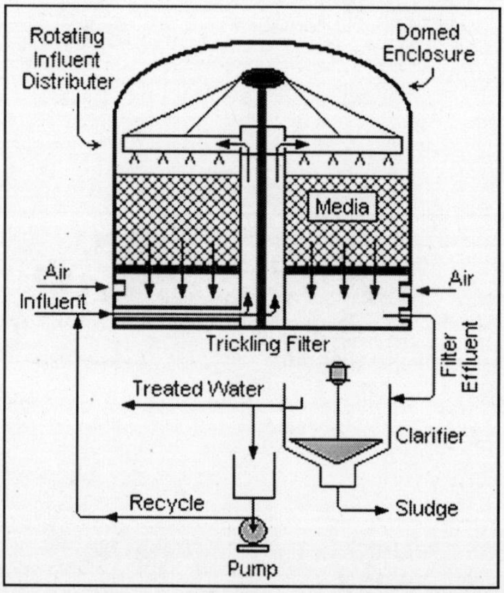

Fig. A Typical Complete Trickling Filter System

A trickling filter consists of a bed of rocks, gravel, slag, peat moss, or plastic media over which wastewater flows downward and contacts a layer (or film) of microbial slime covering the bed media. Aerobic conditions are maintained by forced air flowing through the bed or by natural convection of air. The process involves adsorption of organic compounds in the wastewater by the microbial slime layer, diffusion of air into the slime layer to provide the oxygen required for the biochemical oxidation of the organic compounds. The end products include carbon dioxide gas, water and other products of the oxidation. As the slime layer thickens, it becomes difficult for the air to penetrate the layer and an inner anaerobic layer is formed. The components of a complete trickling filter system are: fundamental components:

- A bed of filter medium upon which a layer of microbial slime is promoted and developed.
- An enclosure or a container which houses the bed of filter medium.
- A system for distributing the flow of wastewater over the filter medium.
- A system for removing and disposing of any sludge from the treated effluent.

The treatment of sewage or other wastewater with trickling filters is among the oldest and most well characterized treatment technologies. A trickling filter is also often called a *trickle filter, trickling biofilter, biofilter, biological filter* or *biological trickling filter*.

TREATMENT OF OTHER ORGANICS

Synthetic organic materials including solvents, paints, pharmaceuticals, pesticides, coking products and so forth can be very difficult to treat. Treatment methods are often specific to the material being treated. Methods include Advanced Oxidation Processing, distillation, adsorption, vitrification, incineration, chemical immobilisation or landfill disposal. Some materials such as some detergents may be capable of biological degradation and in such cases, a modified form of wastewater treatment can be used.

TREATMENT OF ACIDS AND ALKALIS

Acids and alkalis can usually be neutralised under controlled conditions. Neutralisation frequently produces a precipitate that will require treatment as a solid residue that may also be toxic. In some cases, gasses may be evolved requiring treatment for the gas stream. Some other forms of treatment are usually required following neutralisation.

Waste streams rich in hardness ions as from de-ionisation processes can readily loose the hardness ions in a buildup of precipitated calcium and magnesium salts. This precipitation process can cause severe *furring* of pipes and can, in extreme cases, cause the blockage of disposal pipes. A 1 metre diameter industrial marine discharge pipe serving a major chemicals complex was blocked by such salts in the 1970s. Treatment is by concentration of de-ionisation waste waters and disposal to landfill or by careful pH management of the released wastewater.

TREATMENT OF TOXIC MATERIALS

Toxic materials including many organic materials, metals (such as zinc, silver, cadmium, thallium etc.) acids, alkalis, non-metallic elements (such as arsenic or selenium) are generally resistant to biological processes unless very dilute. Metals can often be precipitated out by changing the pH or by treatment with other chemicals. Many, however, are resistant to treatment or mitigation and may require concentration followed by landfilling or recycling. Disolved organics can be *incinerated* within the wastewater by Advanced Oxidation Processes.

INDUSTRIAL WASTE

The Water Corporation's sewer system and wastewater treatment and disposal systems are designed primarily to treat domestic wastewater. However, many businesses and industry generate industrial wastewater streams (or trade waste) which must be disposed of in a cost effective and environmentally responsible manner. The Water Corporation provides an Industrial Waste service to accept sewer discharge of many types of industrial and trade wastes.

This is in recognition of:

- The wider community need for a safe and healthy environment:
- The needs of the business community for cost effective trade waste disposal options: and
- The potential benefits afforded by our municipal wastewater systems.

Of course, there are limits on the types of industrial wastes we can accept. Unauthorised discharges of industrial wastes have the potential to damage our sewer system, disrupt our treatment plant processes, damage the ultimate receiving environment, or pose a hazard to our workers. However, in most instances, our criteria for acceptance are easily addressed with the addition of simple and easily managed pre-treatment fixtures such as grease traps, oil separators, and so on.

To ensure that these environmental and public health controls are maintained, the industrial waste service incorporates a Permit system to manage the totality of industrial wastes entering the public wastewater system. Each business is individually assessed when they connect to the service, and any pre-treatment requirements or other necessary controls on the wastewater discharges from their activity are specified in the conditions associated with their Permit. By complying with their Permit conditions, customers can be assured that their wastes are being properly handled and treated, and can be satisfied that their responsibilities to the community and environment are being met.

The charges for the Industrial Waste service are designed to directly reflect the costs associated with the acceptance and treatment of your industrial wastes into our sewerage systems. These cost reflective charges allow businesses to optimise their use of sewer discharge and other disposal options to choose a solution that is best in terms of both its cost and environmental outcomes. If you are a business intending to discharge wastewater to sewer from kitchens, guest accommodations, workshops, factories, laboratories, or other commercial activity or industrial process, then you are required to apply to the Water Corporation to obtain an Industrial Waste Permit. This is a legal requirement, and is designed to protect against risks to public health, or the environment, which might be caused by toxic substances passing through our sewerage system without our knowledge, or without the proper controls and treatment

THE IMPORTANCE OF INDUSTRIAL WASTE MANAGEMENT

It takes a lot of valuable energy and materials to create and manufacture products and the resulting industrial waste can be difficult to manage. Many cities and countries have put new laws into place to heavily tax companies that produce excess amounts of waste or create potentially harmful effects on the air and ecosystem. The extra taxes help to offset the environment damage by going towards environmental restoration, protection and spreading information to increase knowledge on these issues. People and companies need to educate themselves about the environment. Smog alerts in many cases result from not only harmful transportation emissions but also from the output of factories into the air we breathe. Companies need to be responsible with their industrial waste management and specifically their hazardous waste. Many local governments provide counseling, consulting and recommendations to organizations on what they can do to better manage their waste and plan for a more environmentally friendly production processes. More than ever, there need to be consequences to companies that do not take waste management seriously. Part of this includes reducing harmful emissions into the environment over a period of time and correctly disposing of waste materials.

Countries have terms and conditions about what is acceptable in terms of waste management. Today, more than ever, industries know their impact of manufacturing on smog levels and the escalating cost of managing their

waste. More industrial leaders are showing their accountability for the environment. Citizens need to support companies whose business practices include environmentally conscious and responsible conditions. Using energy more efficiently, reducing the hazardous waste they output into the air and to the landfills and practicing composting and recycling are key factors in improving the way waste is managed.

Companies who have no choice but to continue creating hazardous industrial waste due to the nature of their business need to ensure that they properly dispose of that material and are upfront an honest about the contents of their vehicles, their facilities and management of the waste. Environmental protection acts encourage and reward companies who do their part to more effectively manage waste and work with environmental agencies to maximize efforts to minimizethe impact on the environment.

Industrial waste producers need to pay for the disposal of their materials and in particular, need to take cautionin the way they dispose of hazardous materials. There have been cases documented of companies mislabeling goods and of irresponsible practices leading to contamination of local watersheds. The more that citizens and government push for reform, the more companies will realise that they are accountable for their industrial waste.

WASTEWATER TREATED TO REMOVE POLLUTANTS

Physics, Chemistry, Microbiology and Engineering are all involved in purifying wastewater so that it can be safely returned to the environment. Wastewater treatment plants can be divided into two major types:

BIOLOGICAL AND PHYSICAL/CHEMICAL

- Biological plants are more commonly used to treat domestic or combined domestic and industrial wastewater from a municipality. They use basically the same processes that would occur naturally in the receiving water, but give them a place to happen under controlled conditions, so that the cleansing reactions are completed before the water is discharged into the environment.
- Physical/chemical plants are more often used to treat industrial wastewaters directly, because they often contain pollutants which

cannot be removed efficiently by microorganisms— although industries that deal with biodegradable materials, such as food processing, dairies, breweries, and even paper, plastics and petrochemicals, may use biological treatment. And biological plants generally use some physical and chemical processes, too.

- A physical process usually treats suspended, rather than dissolved pollutants. It may be a *passive* process, such as simply allowing suspended pollutants to settle out or float to the top naturally—depending on whether they are more or less dense than water. Or the process may be aided *mechanically*, such as by gently stirring the water to cause more small particles to bump into each other and stick together, forming larger particles which will settle or rise faster— a process known as*flocculation*. Chemical *flocculants* may also be added to produce larger particles. To aid flotation processes, dissolved air under pressure may be added to cause the formation of tiny bubbles which will attach to particles.

Filtration through a medium such as sand as a final treatment stage can result in a very clear water. *Ultrafiltration, nanofiltration,* and *reverse osmosis* are processes which force water through membranes and can remove *colloidal* material (very fine, electrically charged particles, which will not settle) and even some dissolved matter. Absorption (*adsorption*, technically) on *activated charcoal* is a physical process which can remove dissolved chemicals. *Air or steam stripping*can be used to remove pollutants that are gasses or low-boiling liquids from water, and the vapors which are removed in this way are also often passed through beds of activated charcoal to prevent air pollution. These last processes are used mostly in industrial treatment plants, though activated charcoal is common in municipal plants, as well, for odor control.

Some examples of chemical treatment processes, in an industrial setting, would be:

- Converting a dissolved metal into a solid, settleable form by *precipitation* with an alkaline material like sodium or calcium hydroxide. Dissolved iron or aluminum salts or organic *coagulant* aids like *polyelectrolytes* can be added to help flocculate and settle (or float) the precipitated metal.
- Converting highly toxic cyanides used in mining and metal finishing industries into harmless carbon dioxide and nitrogen by oxidizing them with chlorine
- Destroying organic chemicals by oxidizing them using ozone or hydrogen peroxide, either alone or in combination with catalysts (chemicals which speed up reactions) and/or ultraviolet light

In municipal treatment plants, chemical treatment-- in the form of aluminum or iron salts-- is often used for removal of phosphorus by precipitation. Chlorine or ozone (or ultraviolet light) may be used for disinfection, that is, killing harmful microorganisms before the final discharge of the wastewater. Sulfur dioxide or sulfite solutions can be used to neutralize (reduce) excess chlorine, which is toxic to aquatic life. Chemical coagulants are also used extensively in sludge treatment to thicken the solids and promote the removal of water.

- A typical treatment plant consists of a train of individual unit processes set up in a series, with the output (effluent) of one process becoming the input (influent) of the next process. The first stages will usually be made up of physical processes that take out easily removable pollutants. After this, the remaining pollutants are generally treated further by biological or chemical processes.

These may:

- Convert dissolved or colloidal impurities into a solid or gaseous form, so that they can be removed physically, or
- Convert them into dissolved materials which remain in the water, but are not considered as undesirable as the original pollutants.

The solids (residuals or sludges) which result from these processes form a side stream which also has to be treated for disposal.

A common set of processes that might be found at a municipal treatment plant would be:

- Preliminary treatment to remove large or hard solids that might clog or damage other equipment. These might include grinders, bar screens, and grit channels. The first chops up rags and trash; the second simply catches large objects, which can be raked off; the third allows heavier materials, like sand and stones, to settle out, so that they will not cause abrasive wear on downstream equipment. Grit channels also remove larger food particles (*i.e.*, garbage).
- Primary settling basins, where the water flows slowly for up to a few hours, to allow organic suspended matter to settle out or float to the surface. Most of this material has a density not much different from that of water, so it needs to be given enough time to separate. Settling tanks can be rectangular or circular. In either type, the tank needs to be designed with some type of scrapers at the bottom to collect the settled sludge and direct it to a pit from which it can be pumped for further treatment-- and skimmers at the surface, to collect the material that floats to the top (which is given the rather inglorious name of "scum".) The diagram below shows the operation of a typical primary settling tank.

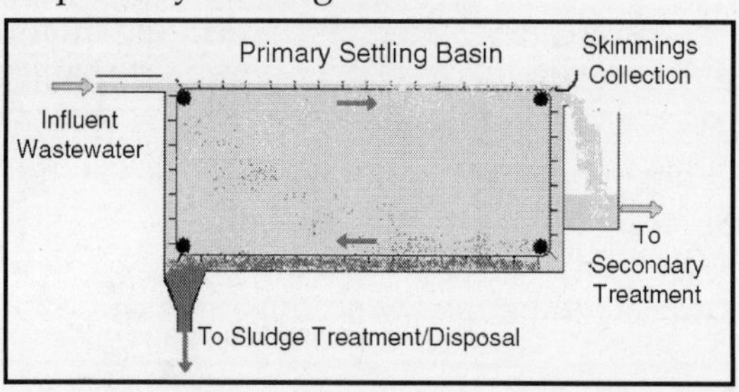

- Secondary treatment, usually biological, tries to remove the remaining dissolved or colloidal organic matter. Generally, the biodegradation of the pollutants is allowed to take place in a location where plenty of air can be supplied to the microorganisms. This promotes formation of the less offensive, oxidized products. Engineers try to design the capacity of the treatment units so that enough of the impurities will be removed to prevent significant oxygen demand in the receiving water after discharge. There are two major types of biological treatment processes: *attached* growth and *suspended* growth. In an attached growth process, the microorganisms grow on a surface, such as rock or plastic.

Examples are:

- Open *trickling filters*, where the water is distributed over rocks and trickles down to underdrains, with air being supplied through vent pipes,
- Enclosed *biotowers*, which are similar, but more likely to use shaped, plastic media instead of rocks, and
- So-called *rotating biological contacters*, or RBC's, which consist of large, partially submerged discs which rotate continuously, so that the microorganisms growing on the disc's surface are repeatedly being exposed alternately to the wastewater and to the air.

The most common type of suspended growth process is the so-called *activated sludge* system. This type of system consists of two parts, an *aeration tank* and a settling tank, or *clarifier*. The aeration tank contains a "sludge" which is what could be best described as a "mixed microbial culture", containing mostly bacteria, as well as protozoa, fungi, algae, etc.

This sludge is constantly mixed and aerated either by compressed air bubblers located along the bottom, or by mechanical aerators on the surface. The wastewater to be treated enters the tank and mixes with the culture, which uses the organic compounds for *growth*— producing more microorganisms— and for *respiration,* which results mostly in the formation of carbon dioxide and water. The process can also be set up to provide biological removal of the nutrients nitrogen and phosphorus.

After sufficient aeration time to reach the required level of treatment, the sludge is carried by the flow into the settling tank, or clarifier, which is often of the circular design. (An important condition for the success of this process is the formation of a type of culture which will flocculate naturally, producing a settling sludge and a reasonably clear upper, or *supernatant* layer.

If the sludge does not behave this way, a lot of solids will be remain in the water leaving the clarifier, and the quality of the effluent wastewater will be poor.) The sludge collected at the bottom of the clarifier is then recycled to the aeration tank to consume more organic material. The term "activated" sludge is used, because by the time the sludge is returned to the aeration tank, the microorganisms have been in an environment depleted of "food" for some time, and are in a "hungry", or activated condition, eager to get busy biodegrading some more wastes. Since the amount of microorganisms, or *biomass*, increases as a result of this process, some must be removed on a regular basis for further treatment and disposal, adding to the solids produced in primary treatment.

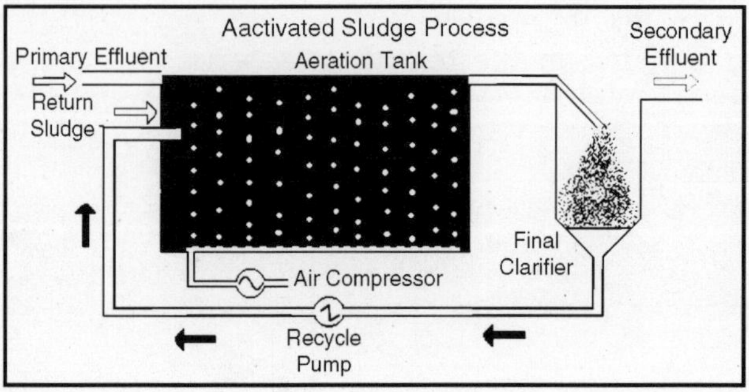

VARIATIONS

Sequencing Batch Reactor (SBR)

The type of activated sludge system described above is a *continuous flow* process. There is a variation in which the entire activated sludge process take place in a *single tank*, but at different times. Steps include filling, aerating, settling, drawing off supernatant, etc. A system like this can provide more flexibility and control over the treatment, including nutrient removal, and is amenable to computer control.

Membrane Bioreactor (MBR)

In this more recent innovation, treated water is pumped out of the aeration tank through banks of microfiltration membranes. Clarifiers are not needed. The sludge concentration can be higher than in a conventional system, which allows treatment in a smaller volume; and the sludge's ability to flocculate well is no longer a consideration. Low effluent solids concentrations can be achieved, which can helps in phosphorus removal and disinfection. for links to some commercial pages which discuss this process. Nutrient removal refers to the treatment of the wastewater to take out nitrogen or phosphorus, which can cause nuisance growth of algae or weeds in the receiving water.

Nitrogen is found in domestic wastewater mostly in the form of ammonia and organic nitrogen. These can be converted to nitrate nitrogen by bacteria, if the plant is designed to provide enough oxygen and a long enough "sludge age" to develop these slow-growing types of organisms. The nitrate which is produced may be discharged; it is still usable as a plant nutrient, but it is much less toxic than ammonia. If more complete removal of nitrogen is required, a biological process can be set up which reduces the nitrate to nitrogen gas (and some nitrous oxide). There are also physical/chemical processes which can remove nitrogen, especially ammonia; they are not as economical for domestic wastewater, but might be suited for an industrial location where no other biological processes are in use. (These methods include alkaline air stripping, ion exchange, and "breakpoint" chlorination.)

Phosphorous removal is most commonly done by chemical precipitation with iron or aluminum compounds, such as ferric chloride or alum (aluminum sulfate). The solids which are produced can be settled along with other sludges, depending on where in the treatment train the process takes place. ("Lime", or calcium hydroxide, also works, but makes the water very alkaline, which has to be corrected, and produces more sludge.). There is also a biological process for phosphorus removal, which depends on designing an activated sludge system in such a way as to promote the development of certain types of bacteria which have the ability to accumulate excess phosphorus within their cells. These methods mainly convert dissolved phosphorus into particulate form. For treatment plants which are required to discharge only very low concentrations of total phosphorus, it is common to have a sand (or other type of) filter as a final stage, to remove most of the suspended solids which may contain phosphorus.

Disinfection, usually the final process before discharge, is the destruction of harmful (*pathogenic*) microorganisms, *i.e.* disease-causing germs. The object is not to kill every living microorganism in the water— which would be *sterilization*— but to reduce the number of harmful ones to levels appropriate for the intended use of the receiving water. The most commonly used disinfectant is chlorine, which can be supplied in the form of a liquefied gas which has to be dissolved in water, or in the form of an alkaline solution called sodium hypochlorite, which is the same compound as common household chlorine bleach. Chlorine is quite effective against most bacteria, but a rather high dose is needed to kill viruses, protozoa, and other forms of pathogen.

Chlorine has several problems associated with its use, among them 1) that it reacts with organic matter to form toxic and carcinogenic chlorinated organics, such as chloroform, 2) chlorine is very toxic to aquatic organisms in the receiving water— the USEPA recommends no more than 0.011 parts per million (mg/L) and 3) it is hazardous to store and handle. Hypochlorite is safer, but still produces problems 1 and 2. Problem 2 can be dealt with by adding sulfur dioxide (liquefied gas) or sodium sulfite or bisulfite (solutions) to neutralize the chlorine. The products are nearly harmless chloride and sulfate ions. This may also help somewhat with problem 1.

A more powerful disinfectant is ozone, an unstable form of oxygen containing three atoms per molecule, rather than the two found in the ordinary oxygen gas which makes up about 21 per cent of the atmosphere. Ozone is too unstable to store, and has to be made as it is used. It is produced by passing an electrical discharge through air, which is then bubbled through the water.

While chlorine can be dosed at a high enough concentration so that some of it remains in the water for a considerable time, ozone is consumed very rapidly and leaves no *residual*. It may also produce some chemical byproducts, but probably not as harmful as those produced by chlorine.

The other commonly used method of disinfection is ultraviolet light. The water is passed through banks of cylindrical, quartz-jacketed fluorescent bulbs. Anything which can absorb the light, such as fouling or scale formation on the bulbs' surfaces, or suspended matter in the water, can interfere with the effectiveness of the disinfection. Some dissolved materials, such as iron and some organic compounds, can also absorb some of the light. Ultraviolet disinfection is becoming more popular because of the increasing complications associated with the use of chlorine.

We have purified and disinfected our wastewater and discharged a clean effluent into the receiving water. Now, what are we going to with all that **sludce** we generated along the way? Sludge from primary settling basins, called primary or "raw" sludge, is a noxious, smelly, gray-black, viscous liquid or semi-solid. It contains very high concentrations of bacteria and other microorganisms, many of them pathogenic, as well as large amounts of biodegradable organic material. Because of the high concentrations, any dissolved oxygen will be consumed rapidly, and the odorous and toxic products of anaerobic biodegradation (*putrefaction*) will be produced. The greasy floatable skimmings from primary treatment are another portion of this *putrescible* solid waste stream.

In addition to the primary sludge, wastewater plants with secondary treatment will produce a "secondary sludge", consisting largely of microorganisms which have grown as a result of consuming the organic wastes. While not quite so objectionable, due to the biodegradation which has already taken place, it is still very high in pathogens and contains much material which will decay and produce odors if not treated further.

Ultimately, the sludge must all be disposed of. The way in which this is done depends on the quality of the sludge— and determines how it needs to be treated. The most desirable final fate for these solids would be for beneficial use in agriculture, since the material has organic matter to act as a soil conditioner, as well as a some fertilizer value. This requires the highest quality "biosolids", free of contamination with toxic metals or industrial organic compounds, and low in pathogens. At a somewhat lower quality, it can be used for similar purposes on non-agricultural land and for land reclamation (*e.g.*, strip mines). Poorer quality sludge can be disposed of by landfilling or incineration. One commonly used method of sludge treatment, called *digestion*, is biological. Since the material is loaded with bacteria and organic matter; why not let the bacteria eat the biodegradable material? Digestion can be either aerobic or anaerobic. *Aerobic* digestion requires supplying oxygen to the sludge; it is similar to the activated sludge process, except no external "food" is provided. In *anaerobic* digestion, the sludge is fed into an air-free vessel; the digestion produces a gas which is mostly a mixture of methane and carbon dioxide. The gas has a fuel value, and can be burned to provide heat to the digester tank and even to run electric generators. Some localities have compressed the gas and used it to power vehicles. Digestion can reduce the amount of organic matter by about 30 to 70 percent, greatly decrease the number of pathogens, and produce a liquid with an inoffensive, "earthy" odor. This makes the sludge safer to dispose of on land, since the odor does not attract as many scavenging pests, such as flies, rodents, gulls, etc., which spread pathogens from the disposal site to other areas— and there are fewer pathogens to be spread.

A liquid sludge, which might contain 3 to 6 per cent dry weight of solids, can be *dewatered* to form a drier sludge *cake* of maybe 15 to 25 percent solids, which can be hauled as a solid rather than having to be handled as a liquid. Equipment used to dewater sludge includes centrifuges, vacuum filters, and belt presses or plate-and-frame presses. Chemical coagulants are commonly added to help form larger aggregates of solids and release the water.

Further processes such as composting and heat drying can produce a drier product with lower pathogen levels. Another approach involves treatment with lime (calcium oxide), which kills pathogens due to its highly alkaline nature as well as the heat that is generated as it reacts with the water in the sludge; this also results in a drier product. A final disposal method which eliminates all of the pathogens and greatly reduces the volume of the sludge is *incineration*. This is not considered a beneficial use, however, and is becoming less popular due to public concerns over air emissions. Sludges from physical-chemical treatment of industrial waste streams containing heavy metals and non-biodegradable toxic organic compounds often must be handled as hazardous wastes. Some of these will end up in hazardous waste landfills, or may be chemically treated for detoxification— or even for recovery of some components for recycling. Recalcitrant organic compounds can be destroyed by carefully controlled high-temperature incineration, or by other innovative processes, such as high-temperature hydrogen reduction.

ODOR CONTROL

In the sewers, prevention of anaerobic conditions and sulfide formation is an important consideration in preventing odors. Hydrogen sulfide is also the major cause of sewer corrosion. If the water is warm and the flow is not rapid enough to aerate the water and scour the pipes, addition of chemicals such as nitrate, hydrogen peroxide or iron compounds can be helpful. In processes where odors are inevitable, the areas must be contained and ventilated. The air then has to be passed through some type of treatment system, such as an activated charcoal bed, a chemical scrubber (often using hypochlorite solution), a compost pile type biofilter— or the air can even be used as part of the air supply for an activated sludge system. I have tried, on this page, to introduce the basics of wastewater treatment. There are many variations and different combinations of treatment processes, as well as innovative methods not discussed here. I have focused on conventional treatment processes, which tend to be built in concrete structures and use a lot of energy-consuming machinery, with the goal of eliminating pollutants. There has been considerable interest, recently, in more "low-tech", environmentally friendly, wastewater treatment processes which consume less energy and allow for recycling of more of the nutrients.

My links page contains some information about some of thesealternative treatment methods. Some companies involved in the production of phosphates for fertilizer use are even interested in recovering these compounds from wastewater to supplement the supply of mined phosphate rock. Water pollution control is a field in which much process research and engineering development is continuing to taking place. Please let me know of any glaring inaccuracies or omissions you may find. (I am a chemist with a wastewater operator's license, not a sanitary or environmental engineer.)

COMPUTERS IN WATER POLLUTION CONTROL

Computers find extensive use in the field of water pollution control. Software stream models are available to predict the effects of pollutants entering the waterways. These can be used by regulatory agencies to set limits on allowable discharges. The movement and treatment of pollutants in groundwater as a result of spills and leaks can be similarly modeled. Programs are available to help deal with the complexities of designing wastewater collection systems (sewers) and predicting the effects of storm water flows. Models are used by engineers to design the types of biological, chemical, and physical wastewater treatment processes described above, so that they will remove pollutants to the required levels.

These programs can also be used by operations staff to optimize the treatment processes, and even for partial automation by accepting real-time information from sensors and using the information to control pumps, valves, aerators, etc. Software is available for scheduling equipment maintenance, tracking plant performance, preparing reports for regulatory agencies, and managing industrial pretreatment programs. High volume analytical laboratories can also use laboratory information management systems (LIMS) to keep track of samples, store analysis and quality control results, and produce reports. In many cases, instruments can be connected directly to the LIMS to reduce transcription errors.

BIOREACTORS SYSTEM FOR WASTE WATER TREATMENT

A bioreactor may refer to any device or system that supports a biologically active environment. In one case, a bioreactor is a vessel in which is carried out a chemical process which involves organisms or biochemically active substances derived from such organisms. This process can either be aerobic or anaerobic. These bioreactors are commonly cylindrical, ranging in size from some liter to cube meters, and are often made of stainless steel.

A bioreactor may also refer to a device or system meant to grow cells or tissues in the context of cell culture. These devices are being developed for use in tissue engineering. On the basis of mode of operation, a bioreactor may be classified as batch, fed batch or continuous. An example of a bioreactor is the chemostat. Organism growing in bioreactor may be suspended or immobilized. The simplest, where cells are immobilized, is a Petri dish with agar gegel.

Large scale immobilized cell bioreactors are:

- Packed bed
- Fibrous bed
- Membrane

Bioreactor Design

Bioreactor design is quite a complex engineering task. Under optimum conditions the microorganisms or cells are able to perform their desired function with great efficiency. The bioreactor's environmental conditions like gas (*i.e.*, air, oxygen, nitrogen, carbon dioxide) flowrates, temperature, pH and dissolved oxygen levels, and agitation speed/circulation rate need to be closely monitored and controlled. Most industrial bioreactor manufacturers use vessels, sensors, controllers, and a control system, networked together for their bioreactor system, see programmable logic controller (PLC).

Fouling can harm the overall sterility and efficiency of the bioreactor, especially the heat exchangers. To avoid it the bioreactor must be easily cleanable and must be as smooth as possible (therefore the round shape). A heat exchanger is needed to maintain the bioprocess at a constant temperature. Biological fermentation is a major source of heat, therefore in most cases bioreactors need water refrigeration. They can be refrigerated with an external jacket or, for very large vessels, with internal coils. In an aerobic process, optimal oxygen transfer is perhaps the most difficult task to accomplish. Oxygen is poorly soluble in water -and even less in fermentation broths- and is relatively scarce in air (20.8 per cent). Oxygen transfer is usually helped by agitation, that is also needed to mix nutrients and to keep the fermentation homogeneous. There are however limits to the speed of agitation, due both to high power consumption (which is proportional to the cube of the speed of the electric motor) and the damage to organisms due to excessive tip speed. Industrial bioreactors usually employ bacteria or other simple organisms that can withstand the forces of agitation.

They are also simple to sustain, requiring only simple nutrient solutions and can grow at astounding rates. In bioreactors where the goal is grow cells or tissues for experimental or therapeutic purposes, the design is significantly different from industrial bioreactors. Many cells and tissues, especially mammalian, must have a surface or other structural support in order to grow, and agitated environments are often destructive to these cell types and tissues. Higher organisms also need more complex growth medium. NASA has developed a new type of bioreactor that artificially grows tissue in cell cultures. NASA's tissue bioreactor can grow heart tissue, skeletal tissue, ligaments, cancer tissue for study, and other types of tissue.

BIOLOGICAL TREATMENT FOR WASTEWATER

Industrial wastewater treatment covers the mechanisms and processes used to treat waters that have been contaminated in some way by man's industrial or commercial activities prior to its release into the environment or its re-use. Most industries produce some wet waste although recent trends in the developed world have been to minimise such production or recycle such waste within the production process. However, many industries remain dependent on processes that produce a water based waste stream.

It is a processes, which employ aerobic or anaerobic microorganisms and result in decanted effluents and separated sludge containing microbial mass together with pollutants. Biological treatment processes are also used in combination and/or in conjunction with mechanical and advanced unit operations. Increasing water pollution and the pathetic state of rivers worldwide demands a wastewater plant that can handle aerobic degradation of waste. Eco-friendliness means, in quintessence, effective pollution control both in the air and on the ground – yet while belching smokestacks are now rarely seen, pollution control on the ground is still a serious cause for concern for the industry, and as a result effluent or wastewater treatment has become mandatory for all industries.

Industrial wastewater treatment is a group of unit processes designed to separate, modify, remove, and destroy undesirable substances carried by wastewater from industrial sources. United States governmental regulations have been issued that involve volatile organic substances, designated priority pollutants; aquatic toxicity as defined by a bioassay; and in some cases nitrogen and phosphorus. As a result, sophisticated technology and process controls have been developed for industrial wastewater treatment. Wastewater streams that are toxic or refractory should be treated at the source, and there are a number of technologies available. For example, wet air oxidation of organic materials at high temperature and pressure (2000 lb/in. or 14 kilopascals and 550°F or 288°C) is restricted to very high concentrations of these substances. Macroreticular (macroporous) resins are specific for the removal of particular organic materials, and the resin is regenerated and used again. Membrane processes, particularly reverse osmosis, are high-pressure operations in which water passes through a semipermeable membrane, leaving the contaminants in a concentrate.

Pretreatment and primary treatment processes address the problems of equalization, neutralization, removal of oil and grease, removal of suspended solids, and precipitation of heavy metals. Aerobic biological treatment is employed for the removal of biodegradable organics. An aerated lagoon system is applicable (where large land areas are available) for treating nontoxic wastewaters, such as generated by pulp and paper mills. Fixed-film processes include the trickling filter and the rotating biological contactor. In these processes, a biofilm is generated on a surface, usually plastic. As the wastewater passes over the film, organics diffuse into the film, where they are biodegraded. Anaerobic processes are sometimes employed before aerobic processes for the treatment of high-strength, readily degradable wastewaters. The primary advantages of the anaerobic process is low sludge production and the generation of energy in the form of methane (CH_4) gas.

Biological processes can remove only degradable organics. Nondegradable organics can be present in the influent wastewater or be generated as oxidation by-products in the biological process. Many of these organics are toxic to aquatic life and must be removed from the effluent before discharge. The most common technology to achieve this objective is adsorption on activated carbon. In some cases, toxic and refractory organics can be pretreated by chemical oxidation using ozone, catalyzed hydrogen peroxide, or advanced oxidation processes. In this case the objective is not mineralization of the organics but detoxification and enhanced biodegradability. Biological nitrogen removal, both nitrification and denitrification, is employed for removal of ammonia from wastewaters. While this process is predictable in the case of municipal wastewaters, many industrial wastewaters are inhibitory to the nitrifying organisms.

Volatile organics can be removed by air or steam stripping. Air stripping is achieved by using packed or tray towers in which air and water counterflow through the tower. In steam stripping, the liquid effluent from the column is separated as an azeotropic mixture. Virtually all of the processes employed for industrial wastewater treatment generate a sludge that requires some means of disposal. In general, the processes employed for thickening and dewatering are the same as those used in municipal wastewater treatment.

Waste activated sludge is usually stabilized by aerobic digestion in which the degradable solids are oxidized by prolonged aeration. Most landfill leachates have high and variable concentrations of organic and inorganic substances. All municipal and most industrial landfill leachates are amenable to biological treatment and can be treated anaerobically or aerobically, depending on the effluent quality desired. Activated carbon has been employed to remove nondegradable organics. In Europe, some plants employ reverse osmosis to produce a high-quality effluent.

AEROBIC TREATMENT SYSTEM

An aerobic treatment system or ATS, often called (incorrectly) an aerobic septic sytem is a small scale sewage treatment system similar to a septic tank system, but which uses an aerobic process for digestion rather than the anaerobic process used in septic systems. These systems are commonly found in rural areas where public sewers are not available, and may be used for a single residence or for a small group of homes.

Process

The ATS process generally consists of the following phases:

- Pre-treatment stage to remove large solids and other undesireable substances from the wastewater.
- Aeration stage, where the aerobic bacteria digest the biological wastes in the wastewater.
- Settling stage to allow any undigested solids to settle. This forms a sludge which must be perioically removed from the system.
- Disinfecting stage, where chlorine or similar disinfectant is mixed with the water, to produce an antiseptic output.

Comparison to Tradional Septic Systems

The aeration state and the disinfecting stage are the primary differences from a traditional septic system to the user of the system. These stages increase the initial cost of the aerobic system, and also the maintainence requirements over the passive septic system. The aerator requires a supply of electricity to drive the air pump, and the disinfectant must be periodically renewed. On the positive side, the aerobic system produces a much higher quality effluent, and the leach field can be far smaller than that of a similar capacity septic system, and the output can be discharged in areas too environmentally sensitive for septic system output. Some aerobic systems recycle the effluent through a sprinkler system, using it to water the lawn.

Aerobic systems are similar to conventional septic systems in that they both use natural processes to treat wastewater. But unlike septic anaerobic treatment, the aerobic treatment process requires oxygen. Aerobic treatment units, therefore, use a mechanism to inject and circulate air inside the treatment tank.

This mechanism requires electricity to operate. For this reason, aerobic systems cost more to operate and need more routine maintenance than most septic systems. However, when properly operated and maintained, aerobic systems can provide a high quality wastewater treatment alternative to septic systems.

Why Use Aerobic Treatment

Aerobic Treatment Units (ATUs) have proven to be a reliable technology when they are properly designed, constructed, and maintained. When properly monitored and maintained their performance is consistent. ATUs may enable development or use of difficult sites. They can remedy existing malfunctioning systems and they can be a good option for homes in environmentally sensitive areas.

Process Description

Aerobic systems treat wastewater using natural processes that require oxygen. Bacteria that thrive in oxygen-rich environments work to break down and digest the wastewater inside the aerobic treatment unit. Like most onsite systems, aerobic systems treat the wastewater in stages. Sometimes the wastewater receives pretreatment before it enters the aerobic unit, and the treated wastewater leaving the unit requires additional treatment or disinfection before being returned to the environment. Such a variety of designs exists for home aerobic units and systems that it is impossible to describe a typical system. Instead, it is more practical to discuss how some common design features of aerobic systems work and the different stages of aerobic treatment.

Pretreatment

Some aerobic systems include a pretreatment step to reduce the amount of solids in the wastewater going into the aerobic unit. Solids include greases, oils, toilet paper, and other materials that are put down the drain or flushed into the system. Too much solid material can clog the unit and prevent effective treatment. Some pretreatment methods include a septic tank, a primary settling compartment in the pretreatment unit, or a trash trap. Pretreatment is optional but can greatly improve a unit's performance.

Aerobic Treatment Units

The main function of the aerobic treatment unit is to collect and treat household wastewater, which includes all water from toilets, bathtubs, showers, sinks, and laundry. Aerobic units themselves come in many sizes and shapes-rectangular, conical, and some shapes that defy classification.

There are two typical aerobic treatment designs:

1. Suspended growth units and
2. Attached growth units.

The process most aerobic units use to treat wastewater is referred to as suspended growth. These units include a main compartment called an aeration chamber in which air is mixed with the wastewater. Since most home aerobic units are buried underground like septic tanks, the air must be forced into the aeration chamber by an air blower or through liquid agitation. The forced air mixes with wastewater in the aeration chamber, and the oxygen supports the growth of aerobic bacterial that digests the solids in the wastewater. This mixture of wastewater and oxygen is called the mixed liquor.

The treatment occurring in the mixed liquor is referred to as suspended growth because the bacterial grow as they are suspended in the liquid unattached to any surface. Unfortunately, the bacterial cannot digest all of the solids in the mixed liquor, and these solids eventually settle out as sludge. Many aerobic units include a secondary chamber called a settling chamber or clarifier where excess solids can settle. Other designs allow the sludge to accumulate at the bottom of the tank. In aerobic units designed with a separate settling compartment, the sludge returns to the aeration chamber (either by gravity or by a pumping device). The sludge contains bacterial that also aid in the treatment process.

Although, in theory, the aerobic treatment process should eventually be able to consume the sludge completely, in practice, the sludge does build up in most units and will need to be pumped out periodically so that solids don't clog the unit. An alternative design for aerobic treatment is the attached growth system. These units treat wastewater by taking a surface made of material that the bacteria can attach to, and then exposing that surface alternately to wastewater and air. This is done either by rotating the surface in and out of the wastewater or by dosing the wastewater onto the surface. Pretreatment is required. The air needed for the process either is naturally present or is supplied mechanically.

Attached growth systems, such as trickling filters and rotating disks, are less common than suspended growth systems, but have certain advantages. For example, there is no need for mixing, and solids are less likely to be washed out of the system during periods of heavy household water use.

Flow Design

The way and the rate in which wastewater is received by and flows through the aerobic unit differs from design to design. Continuous flow designs simply allow the wastewater to flow through the unit at the same rate that it leaves the home. Other designs employ devices (such as pretreatment tanks, surge chambers, and baffles) to control the amount of the incoming flow. Batch process designs use pumps or siphons to control the amount of wastewater in the aeration tank and/or to discharge the treated wastewater in controlled amounts after a certain period.

Controlling the flow of wastewater helps to protect the treatment process. When too much wastewater is flushed into the system all at once, it can become overburdened, and the quality of treatment can suffer. The disadvantages to mechanical flow control devices are that, like all mechanical components, they need maintenance and run the risk of malfunctioning.

Disinfection

Some units have the disinfection process incorporated into the unit design. In some cases, disinfection may be the only treatment required of the wastewater from an aerobic unit before the water is released into the environment. Added costs for disinfectants, such as chlorine, should be taken into account with aerobic units.

Design Criteria

Aerobic units should be large enough to allow enough time for the solids to settle and for the wastewater to be treated. The daily wastewater volume is usually determined by the number of bedrooms in the house. Flows in Arizona for individual homes are up to 140 gallons per day (gpd) per bedroom. The needed size of an aerobic unit is often estimated the same way the size of a septic tank is estimated, by the number of bedrooms (not bathrooms) in the house. It is assumed that each person will use approximately 50 to 100 gallons of water per day, and that each bedroom can accommodate two people. When calculated this way, a three-bedroom house will require a unit with a capacity of 300 to 600 gallons per day.

Some health departments require that aerobic units be sized at least as large as a septic tank in case the aerobic unit malfunctions and oxygen doesn't mix with the wastewater. In such cases, the aerobic unit will work as a septic tank, which will, at least, provide partial treatment for the wastewater. Lower temperatures tend to slow down most biological processes, and higher temperatures tend to speed them up. The aerobic process itself creates heat, which, along with the heat from the electrical components, may help to keep the treatment process active. However, cold weather can have adverse effects on the performance of aerobic units.

In one 1977 study of aerobic units, bulking of the sludge seemed to occur when the temperature of the mixed liquor fell below 15 degrees Celsius (59 degrees Fahrenheit). Problems can sometimes be avoided by insulating around the units.

Operation and Maintenance

Aerobic treatment systems are not accepted in all areas. Regulations for onsite systems can vary from state to state and county to county. A major reason that aerobic systems are not more widely used is concern about improper operation and maintenance by homeowners. Aerobic systems require regular maintenance, and abuse or neglect can easily lead to system failure. A typical inspection should include removing the unit's cover and checking its general appearance. Check pipes and the inside of the aeration chamber and note the appearance of the wastewater inside the unit and its color and odor. If the unit includes a chlorinator, this too will need to be checked and may need cleaning. Samples may be taken of the mixed liquor from the aeration chamber, as well as the final treated wastewater. Check to see that all mechanical parts, alarms, and controls are in working order, and that solids are pumped from the system if needed.

It is important that mechanical components in aerobic systems receive regular inspection and maintenance. For example, air compressors sometimes need to be oiled, and vanes, filters, and seals may need to be replaced. Malfunctions are common during the first few months after installation. In most cases, homeowners do not have the expertise to inspect, repair, and maintain their own systems.

Most aerobic units have controls that can be switched on and off by the homeowner in case of emergency. Aerobic units also are required to have alarms to alert the homeowner of malfunctions. Depending on the design of the system, controls and alarms can be located either inside or outside the home, and alarms can be visible, audible, or both. Homeowners should make sure that controls and alarms are always protected from corrosion, and that the aerobic unit is turned back on if here is a power outage or if it is turned off temporarily.

To assure homeowners are receiving a reputable ATU most states require the system to by approved by the NSF International (formerly the National Sanitation Foundation). NSF has tested aerobic units according to the requirements of ANSI/NSF Standard 40. NSF is a nonprofit organization devoted to the protection of the environment through the development of product standards, product evaluations, research, education and training. The American National Standards Institute (ANSI) is the recognized accredited in the U.S. for organizations that develop consumer standards and for those that provide independent product evaluations. NSF is accredited by ANSI for both of these areas of service.

ANAEROBIC DIGESTION

Anaerobic digestion (AD) is the harnessed and contained, naturally occurring process of anaerobic decomposition. An anaerobic digester is an industrial system that harnesses these natural process to treat waste, produce biogas that can be used to power electricity generators, provide heat and produce soil improving material. Anaerobic digesters have been around for a long time and they are commonly used for sewage treatment or for managing animal waste. Increasing environmental pressures on waste disposal has increased the use of AD as a process for reducing waste volumes and generating useful byproducts. It is a fairly simple process that can greatly reduce the amount of organic matter which might otherwise end up in landfills or waste incinerators.

Almost any organic material can be processed in this manner. This includes biodegradable waste materials such as waste paper, grass clippings, leftover food, sewage and animal waste. Anaerobic digesters can also be fed with specially grown energy crops to boost biodegradable content and hence increase biogas production. After sorting or screening to remove inorganic or hazardous materials such as metals and plastics, the material to be processed is often shredded, minced, or hydrocrushed to increase the surface area available to microbes in the digesters and hence increase the speed of digestion. The material is then fed into an airtight digester often with extra water added depending on the digestion process and feedstock.

MICROBIAL FUEL CELL

A microbial fuel cell (MFC) or biological fuel cell is a device in which a chemical, typically glucose, is converted to electric power by means of bacteria on the anode side. Power outputs are usually small, in the order of magnitude of about a milliwatt, and there are no current commercially available applications. However, some hope to use them in the future to build a glucose-powered pacemakers that would need no other power supply than the glucose present in the bloodstream.

A microbial Fuel Cell is a device that converts chemical energy to electrical energy by the catalytic reaction of microorganisms. A typical microbial fuel cell consists of anode and cathode compartments separated by a cation specific membrane. In the anode compartment, fuel is oxidized by microorganisms, generating electrons and protons. Electrons are transferred to the cathode compartment through, and the protons through the membrane. Electrons and protons are consumed in the cathode compartment reducing oxygen to water. In general, there are two types of microbial fuel cell, mediator- and mediator-less microbial fuel cell. Biological fuel cells take glucose and methanol from food scraps and convert it into hydrogen and food for the bacteria.

Mediator Microbial Fuel Cell

Most of the microbial cells are electrochemically inactive. The electron transfer from microbial cells to the electrode is facilitated by mediators such as potassium ferric cyanide, thionine, methyl viologen (methyl blue), humic acid, neutral red and so on (Delaney *et al.*, 1984; Lithgow *et al.*, 1986). Most of the mediators available are expensive and toxic.

Mediator-less Microbial Fuel Cell

Mediator-less microbial fuel cells have been engineered at the Korea Institute of Science and Technology, by a team led by Kim, Byung Hong A mediator-less microbial fuel cell does not require a mediator but uses electrochemically active bacteria to transfer electrons to the electrode (electrons are carried directly from the bacterial respiratory enzyme to the electrode). Among the electrochemically active bacteria are, *Shewanella putrefaciens* (Kim *et al.*, 1999a), *Aeromonas hydrophila* (Cuong *et al.*, 2003), and others.

Mediator-less MFCs are a much more recent development and due to this the factors that affect optimum operation, such as the bacteria used in the system, the type of ion membrane, and the system conditions such as temperature, are not particularly well understood. Bacteria in mediator-less MFCs typically have electrochemically-active redox enzymes such as cytochromes on their outer membrane that can transfer electrons to external materials.

WASTE TREATMENT TECHNOLOGY

Waste treatment refers to the activities required to ensure that waste has the least practicable impact on the environment. In many countries various forms of waste treatment are required by law. There are a variety of waste treatment technologies. The long established method of waste treatment is simply landfill. Other technologies include composting and recycling. There are also a number of new and emerging waste treatment technologies which are able to treat a variety of waste streams.

There is a move away from treating waste as a disposal problem towards recognising waste contains valuable resources that can be recovered. Recovery of value from waste can either be in the form of recovery of resources or recovery of energy. Environmental biotechnology is the use of living organisms for a wide variety of applications in hazardous waste treatment and pollution control. Marine biotechnologists are studying ways that estuarine bacteria can detoxify materials such as chemical sea brines that cause environmental problems in many industries.

Biotechnology in industry employs the techniques of modern molecular biology to reduce the environmental impact of manufacturing. Industrial biotechnology also works to make manufacturing processes more efficient for industries such as textiles, paper and pulp, and specialty chemicals. Biotechnology is improving the safety of raw materials by helping food scientists discover the exact identity of the allergenic protein in foods such as peanuts, soybeans and milk, so they can then remove them. Biotechnology has provided law enforcement professionals with another way of placing a suspect at the scene of a crime. This area of study, called forensic biotechnology, uses a method called DNA fingerprinting.

BIOLOGICAL TREATMENT METHOD

Glial cell line-derived neurotrophic factor (GDNF), a distant member of the transforming growth factor-â (TGFâ) family, is a protein that is essential for the survival of dopaminergic, motor, and peripheral neurons. To facilitate its study, biotechnology has developed sensitive (low pg/ml) enzyme-linked immunosorbant assays (ELISA) to quantitate endogenous concentrations of GDNF, along with neurotrophin-3 (NT-3) and nerve growth factor (NGF). However, endogenous tissue levels of GDNF in adult animals are not readily detected by ELISA and do not correlate well with message RNA. Based upon previously described methods for the extraction of TGFâ from tissue samples, we have developed an acid-treatment procedure to allow the quantification of total endogenous GDNF.

This procedure also was evaluated for use when measuring total endogenous levels of NT-3 and NGF from biological samples. The acid-treatment procedure increases the detectable amounts of GDNF, NT-3, and NGF in all tissue samples and most of the serum samples tested. Moreover, these values were as much as 35 times greater than those detected using traditional extraction buffers. Such elevated concentrations likely resulted from the acid treatment promoting the dissociation of ligands from receptors or binding proteins, thereby making more of the analyte available to be measured in the ELISA These findings indicate that appropriate sample treatment is essential for the measurement of total endogenous neurotrophic factors. The intensification of agriculture and food production in recent years has led to an increase in the production of food co-products and wastes. Their disposal by incineration or landfill is often expensive as well as environmentally sensitive. Methods to valorise unused co-products and improve the management of wastes that cannot be reused, as well as techniques to reduce the quantity of waste produced in the first place, are increasingly important to the food industry.

INDUSTRIAL COMPOSTING

Industrial composting systems are being increasingly installed as an alternative form of waste management to landfill along with other advanced waste processing systems. The industrial composting or anaerobic digestion can be combined with mechanical sorting of mixed waste streams and is given the term mechanical biological treatment. Industrial composting helps prevent global warming by treatment of biodegradable waste before it enters landfill. Once this waste is landfilled it breaks down anaerobically producing landfill gas that contains methane, a potent greenhouse gas.

Most commercial and industrial composting operations use active composting techniques. This ensures a higher quality product and produces results in the shortest time (see compost windrow turner). The greatest control, and therefore the highest quality, is generally achieved by composting inside an enclosed vessel which is monitored and adjusted continuously for optimal temperature, air flow, moisture, and other parameters. Large scale composting systems are used by a few urban centers around the world. Co-composting is a technique which combines solid waste with de-watered bio-solids. The world's largest co-composter is in Edmonton, Alberta, Canada, which turns 220,000 tonnes of residential solid waste and 22,500 dry tonnes of biosolids per year into 80,000 tonnes of compost using a facility 38,690 square metres in size (equivalent to 8 football fields).

Biodegradable Waste Management

Anaerobic digestion breaks down the biodegradable component of the waste to produce biogas and soil improver. The biogas can be used to generate renewable energy. Some processes through water based processing enable high rates of gas and green energy generation without the production of RDF. This is facilitated by processing the waste in water.

Biological can also refer to a composting stage. Here the organic component is treated with aerobic microorganisms. They break down the waste into carbon dioxide and compost. There is no green energy produced by systems employing only composting treatment for the biodegradable waste. Some systems, such as UR-3R involve both a partial anaerobic digestion stage along with a secondary composting stage. By processing the biodegradable waste either by anaerobic digestion or by composting MBT technologies help to reduce the contribution of greenhouse gases to global warming.

Bioaccumulation

To bioaccumulate literally means to accumulate in a biological system. However, it is commonly taken to measure the uptake over time of a substance, called a bioaccumulant, that can accumulate in a biological system. Bioaccumulation can be divided into bioconcentration and biomagnification. Bioconcentration considers uptake from the non-living environment while biomagnification describes uptake through the food chain. For many fat-soluble and persistent chemicals (POPs), biomagnification is the dominant factor.

Everything in a biological system has a biological halflife, that is, a measure of how long it will stay in that system until it is lost, is excreted, degrades, reacts into something different, or ends its presence in some other way. Most substances have a short half-life, as they are metabolized, or excreted as waste. However, some compounds may stay in a system for a much longer period of time. For example, calcium in the human body is laid down in bones and teeth, and even when bone cells die, their calcium is used again in the building of bones. This is a sensible and efficient re-use of scarce resources. The problem arises when toxic substances stay in the body for a long period of time. They are not acutely poisonous, otherwise they would kill straight away, but are associated with chronic poisoning. If the input of a toxic substance to an organism is greater than the rate at which the substance is lost, the organism is said to be bioaccumulating that substance. Thus, the longer the biological half-life of the substance the greater the risk of chronic poisoning, even if environmental levels of the toxin are very low. This is one reason why chronic poisoning is a common aspect of environmental health in the workplace. As people spend so much time, for so many years in these environments, very low levels of toxins can be lethal over time.

An example of poisoning in the workplace can be seen from the phrase "as mad as a hatter". The process for stiffening the felt used in making hats involved mercury, which forms organic species such as methylmercury, which is lipid soluble, and tends to accumulate in the brain resulting in mercury poisoning. Other lipid (fat) soluble poisons include tetra-ethyl lead compounds (the lead in leaded petrol), and DDT. These compounds are stored in the body's fat, and in times of famine, when the fatty tissues are used for energy, the compounds are released and cause acute poisoning.

Strontium 90, part of the fallout from atomic bombs, is mistaken by the human body for calcium, and is laid down in the bone, where its radiation can cause damage for a long time. Naturally produced toxins can also bioaccumulate. The marine algal blooms known as "red tides" can result in local filter feeding organisms such as mussels and oysters becoming toxic; coral fish can be responsible for the poisoning known as ciguatera when they accumulate a toxin called ciguatoxin from reef algae. Other compounds that are not normally considered toxic can be accumulated to toxic levels in organisms. The classic example is of Vitamin A, which becomes concentrated in carnivore livers of *e.g.* polar bears: as a pure carnivore that feeds on other carnivores (seals), they accumulate extremely large amounts of Vitamin A in their livers. It was known by the native peoples of the arctic that the livers should not be eaten, but arctic explorers have suffered Hypervitaminosis A from eating the bear livers (and there has been at least one example of similar poisoning of Antarctic explorers eating husky dog livers).

4

Sewage Treatment

Sewage treatment is the process of removing contaminants from wastewater and household sewage, both runoff (effluents) and domestic. It includes physical, chemical, and biological processes to remove physical, chemical and biological contaminants. Its objective is to produce an environmentally safe fluid waste stream (or treated effluent) and a solid waste (or treated sludge) suitable for disposal or reuse (usually as farm fertilizer). Using advanced technology it is now possible to re-use sewage effluent for drinking water, although Singapore is the only country to implement such technology on a production scale in its production of NEWater.

Sewage treatment, or domestic wastewater treatment, is the process of removing contaminants from wastewater, both runoff (effluents) and domestic. It includes physical, chemical and biological processes to remove physical, chemical and biological contaminants. Its objective is to produce a waste stream (or treated effluent) and a solid waste or sludge suitable for discharge or reuse back into the environment. This material is often inadvertently contaminated with many toxic organic and inorganic compounds. Sewage is created by residences, institutions, hospitals and commercial and industrial establishments. It can be treated close to where it is created (in septic tanks, biofilters or aerobic treatment systems), or collected and transported via a network of pipes and pump stations to a municipal treatment plant. Sewage collection and treatment is typically subject to local, state and federal regulations and standards. Industrial sources of wastewater often require specialized treatment processes.

The sewage treatment involves three stages, called *primary*, *secondary* and *tertiary treatment*. First, the solids are separated from the wastewater stream. Then dissolved biological matter is progressively converted into a solid mass by using indigenous, water-borne microorganisms. Finally, the biological solids are neutralized then disposed of or re-used, and the treated water may be disinfected chemically or physically (for example by lagoons and micro-filtration). The final effluent can be discharged into a stream, river, bay, lagoon or wetland, or it can be used for the irrigation of a golf course, green way or park. If it is sufficiently clean, it can also be used for groundwater recharge.

DESCRIPTION

Raw influent (sewage) includes household waste liquid from toilets, baths, showers, kitchens, sinks, and so forth that is disposed of via sewers. In many areas, sewage also includes liquid waste from industry and commerce. The draining of household waste into greywater and blackwater is becoming more common in the developed world, with greywater being permitted to be used for watering plants or recycled for flushing toilets. A lot of sewage also includes some surface water from roofs or hard-standing areas. Municipal wastewater therefore includes residential, commercial, and industrial liquid waste discharges, and may include stormwater runoff.

Sewage systems capable of handling stormwater are known as combined systems or combined sewers. Such systems are usually avoided since they complicate and thereby reduce the efficiency of sewage treatment plants owing to their seasonality. The variability in flow also leads to often larger than necessary, and subsequently more expensive, treatment facilities. In addition, heavy storms that contribute more flows than the treatment plant can handle may overwhelm the sewage treatment system, causing a spill or overflow (called a combined sewer overflow, or CSO, in the United States). It is preferable to have a separate storm drain system for stormwater in areas that are developed with sewer systems.

As rainfall runs over the surface of roofs and the ground, it may pick up various contaminants including soil particles and other sediment, heavy metals, organic compounds, animal waste, and oil and grease. Some jurisdictions require stormwater to receive some level of treatment before being discharged directly into waterways. Examples of treatment processes used for stormwater include sedimentation basins, wetlands, buried concrete vaults with various kinds of filters, and vortex separators (to remove coarse solids). The site where the raw wastewater is processed before it is discharged back to the environment is called a wastewater treatment plant (WWTP). The order and types of mechanical, chemical and biological systems that comprise the wastewater treatment plant are typically the same for most developed countries:

- Mechanical treatment
- Influx (Influent)
- Removal of large objects
- Removal of sand and grit
- Pre-precipitation
- Biological treatment
- Oxidation bed (oxidizing bed) or aeration system
- Post precipitation
- Chemical treatment (this step is usually combined with settling and other processes to remove solids, such as filtration. The combination is referred to in the U.S. as physic

Primary treatment removes the materials that can be easily collected from the raw wastewater and disposed of. The typical materials that are removed during primary treatment include fats, oils, and greases (also referred to as FOG), sand, gravels and rocks (also referred to as grit), larger settleable solids and floating materials (such as rags and flushed feminine hygiene products). This step is done entirely with machinery.

CHARACTERISTICS OF SEWAGE SLUDGE

Most wastewater treatment processes produce a sludge which has to be disposed of. Conventional secondary sewage treatment plants typically generate a primary sludge in the primary sedimentation stage of treatment and a secondary, biological, sludge in final sedimentation after the biological process. The characteristics of the secondary sludge vary with the type of biological process and, often, it is mixed with primary sludge before treatment and disposal. Approximately one half of the costs of operating secondary sewage treatment plants in Europe can be associated with sludge treatment and disposal.

Land application of raw or treated sewage sludge can reduce significantly the sludge disposal cost component of sewage treatment as well as providing a large part of the nitrogen and phosphorus requirements of many crops. Very rarely do urban sewerage systems transport only domestic sewage to treatment plants; industrial effluents and storm-water runoff from roads and other paved areas are frequently discharged into sewers. Thus sewage sludge will contain, in addition to organic waste material, traces of many pollutants used in our modern society. Some of these substances can be phytotoxic and some toxic to humans and/or animals so it is necessary to control the concentrations in the soil of potentially toxic elements and their rate of application to the soil. The risk to health of chemicals in sewage sludge applied to land has been reviewed by Dean and Suess.

Sewage sludge also contains pathogenic bacteria, viruses and protozoa along with other parasitic helminths which can give rise to potential hazards to the health of humans, animals and plants. A WHO Report on the risk to health of microbes in sewage sludge applied to land identified salmonellae and *Taenia* as giving rise to greatest concern. The numbers of pathogenic and parasitic organisms in sludge can be significantly reduced before application to the land by appropriate sludge treatment and the potential health risk is further reduced by the effects of climate, soil-microorganisms and time after the sludge is applied to the soil. Nevertheless, in the case of certain crops, limitations on planting, grazing and harvesting are necessary. Apart from those components of concern, sewage sludge also contains useful concentrations of nitrogen, phosphorus and organic matter. The availability of the phosphorus content in the year of application is about 50 per cent and is independent of any prior sludge treatment. Nitrogen availability is more dependent on sludge treatment, untreated liquid sludge and dewatered treated sludge releasing nitrogen slowly with the benefits to crops being realised over a relatively long period.

Liquid anaerobically-digested sludge has high ammonia-nitrogen content which is readily available to plants and can be of particular benefit to grassland. The organic matter in sludge can improve the water retaining capacity and structure of some soils, especially when applied in the form of dewatered sludge cake. Directive prohibits the sludge from sewage treatment plants from being used in agriculture unless specified requirements are fulfilled, including the testing of the sludge and the soil.

Parameters subject to the provisions of the Directive include the following:

- Dry matter (per cent)
- Organic matter (per cent dry solids)
- Copper (mg/kg dry solids)
- Nickel (mg/kg dry solids)
- pH
- Nitrogen, total and ammoniacal (per cent dry solids)
- Phosphorus, total (per cent dry solids)
- Zinc (mg/kg dry solids)
- Cadmium (mg/kg dry solids)
- Lead (mg/kg dry solids)
- Mercury (mg/kg dry solids)
- Chromium (mg/kg dry solids)

To these parameters the UK Department of the Environment has added molybdenum, selenium, arsenic and fluoride in the recent 'Code of Practice for Agricultural Use of Sewage Sludge'. Sludge must be analyzed for the Directive parameters at least once every 6 months and every time significant changes occur in the quality of the sewage treated. The frequency of analysis for the additional four parameters may be reduced to not less than once in five years provided that their concentrations in the sludge are consistently no greater than the following reference concentrations: Mb - 3mg/kg dry solids, Se - 2mg/kg dry solids, As -2mg/kg dry solids and Fl - 200mg/kg dry solids.

COLLECTION OF SEWAGE

House drains, that is to say, those parts of the domestic system of drainage which extend from the soil-pipes and waste-pipes to the sewer, are generally made of glazed stoneware pipes having a diameter of 4 in., 6 in., or sometimes 9 or 12 in., according to the estimated amount of waste to be removed. In anent ordinary domestic dwellings there is rarely any occasion to use pipes of a greater diameter than 6 in., and this only for the main drain, the branches and single lines of piping being 4 in. in diameter. It is a good rule to make the pipes and other fittings, such as channels and bends, as small in diameter as possible, having due regard to efficient capacity.

Such a drain is more cleanly than one too large for its purpose, in that it is more thoroughly flushed when in use, the sewage running at a much faster speed through a full pipe than through one only partially full. For this reason a pipe having too great a capacity for the work it has to do is liable to become corroded by sediment deposited from slowly moving waste.

The pipes are made in 2 ft. lengths and are formed with a socket at one end into which the straight end of the next pipe fits loosely. This is wedged in position with a little gasket and the remaining space then carefully filled with neat Portland cement. Pipes are made also with a bituminous substance in the socket and around the spigot end, and by merely pushing the one into the other the joint is made.

The bitumen is curved to allow self-adjustment to any slight settlement, so that damage to the joint is avoided. A composite joint may be used having the bitumen lining reinforced with the ordinary Portland cement filling. This type is somewhat more expensive than the ordinary jointing, but it makes a powerful and effective connexion. The method of connecting two lead pipes by a " wiped solder joint " the method of connect amen bihimatn ing a lead pipe into the socket of a stoneware one, a brass sleeve piece or ferrule being used to give the necessary stiffness to the end of the lead pipe. This arrangement is frequently used, for example, at the base of a soil-pipe at its junction with the drain. The lead pipe has a brass socket attached to it to take the plain end of a stoneware pipe. This form of connexion is used between a water-closet and a lead trap.

In the water-carriage system of drainage each house has its own network of drain-pipes laid under the ground, into which are taken the waste-pipes which lead from the closets, urinals, sinks, lavatory basins, and rain-water and other gulleys within and about the house. The many branches are gathered into one or more manholes, and connexion is finally made by means of a single pipe with the common public sewer. Gas from the sewer is prevented from entering the house drains by a disconnecting trap fixed in the manhole nearest the entrance to the sewer.

The fundamental maxims of house sanitation are first, that there shall be complete disconnexion between the pipes within and without the house, and second, that the drainage shall be so constructed as to allow for the free admission of air in order to secure the thorough ventilation of all parts of the system and avoid the possibility of the accumulation of gas in any of the wasteor drain-pipes.

The drains must be planned to conduct the waste material from the premises as quickly as possible without leakage or deposit by the way. The pipes should be laid in straight lines from point to point to true gradients of between 2 to 4 in. in io ft. Junctions with branch pipes and any bends necessary should be gathered, as far as practicable, in inspection chambers fitted with open channels instead of closed pipes.

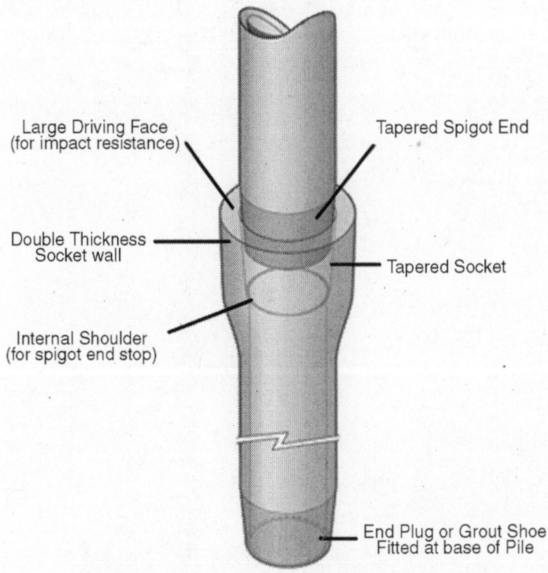

Fig. Iron Spigot and Socket Joint.

This allows of easy inspection and testing, and provides means of access for the drain-rods in cases of blockage. Sometimes it is desired, for reasons of economy or otherwise, to avoid the use of a manhole at a change of direction in the drain. A branch pipe which may have a specially shaped junction for cleaning the pipes in both directions is taken up with a slope to the ground or floor level and there finished with an air-tight cover which may be removed to allow the introduction of drain-rods should the pipes become blocked.

Junctions of one another should be made obliquely in the direction of the floor. Stoneware pipes should be laid upon a bed of concrete not less than 6 in. thick and benched up at the sides with concrete to prevent any movement. When such pipes pass under a building they should be entirely surrounded by a concrete casing at least 6 in. in thickness. No drain should lie under a building if it is possible to avoid it, for injury is very liable to occur through some slight settlement of the building, and in a position such that the smells escaping from the damaged pipe would rise up through the floor into the building this would be an especially serious matter.

The expense and annoyance of having the ground opened up for the repair of defects in the pipes beneath is another strong argument against drains being placed under a house. Where this is really necessary, however, pipes of cast iron are recommended instead of the ordinary stoneware pipes, as being stronger; being made in lengths of 6 and 9 ft., they have a great advantage over the 2 ft. long stoneware tubes, for the joints of the latter are frequently a source of weakness. The joints, fewer in number, are made with molten lead, or flanged pipes are used and the joints packed with bolted.

The principle of disconnexion and outdoor pipes should be retained between the latter and the sewer, and the domestic system should be cut off from the public drain by means of a dis connecting trap. This appliance is usually placed in a small chamber or manhole, easy of access for inspection, built close to the boundary of the premises, and as near as possible to the sewer into which the house drain discharges.

Fig. Iron-flanged.

The cap to the clearing arm has a chain attached by which it can be removed in case of flooding. The channels are benched up at the sides with cement, and the manhole is rendered on the inside with a cement lining. A fresh air inlet is taken out near the top of the chamber and is fitted with a mica flap inlet valve. The cover is of cast-iron in a cast-iron frame shaped with grooves to afford a double seal, the grooves being filled with a composition of tallow and fine sand.

Where there is a danger of a backflow from the sewer due to its becoming flooded, a hinged flap should be placed at the junction of drain and sewer to prevent sewage from entering the house drain. A ball trap designe-i for this purpose may be used in place of a flap, and is more satisfactory, for the latter is liable to become corroded and work stiffly. In the ball-trap appliance the flowing back of the sewage forces a copper ball to fit tightly against the drain outlet, the ball dropping out of the way of the flow directly the pressure is relaxed. The water-carriage system of drainage is undoubtedly the most nearly perfect yet devised. At the same time it is a very costly system to install with its network of sewers, stations, and arrangements for depositing pumping ? g P g the sewage either in the sea or river, or upon the land or " sewage farm." In country districts and small towns and villages, however, excreta are often collected in small vessels Section. The dry-earth system oxidizing effect which a porous substance such as dry earth exerts by bringing any sewage with which it is mixed into intimate contact with the air contained in its pores. The system is of rather limited application from the fact that it leaves other constituents of sewage to be dealt with by other means. But so far as it goes it is excellent, and where there is no general system of watercarriage sewerage an earth-closet will in careful hands give perfect satisfaction.

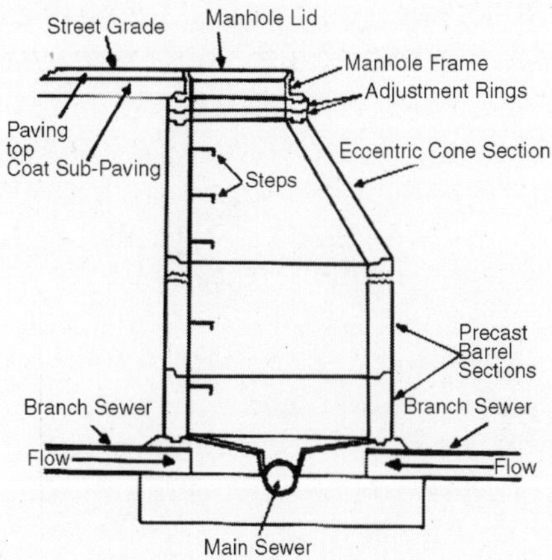

Fig. Manhole.

Numerous forms of earth-closet are sold in which a suitable quantity of earth is automatically thrown into the pan at each time of use but a box filled with dry earth and a hand scoop will answer the purpose nearly as well. A plan much used in towns on the continent of Europe 24 is to collect excrement in air-tight vaults which are emptied at intervals into a tank cart by a suction pump. Another pneumatic system adopted on the continent has the cesspools at individual houses per manently connected with a central reservoir by pipes through which the contents of the former are sucked by exhausting air from the reservoir at the central station.

Newly laid drains should be carefully tested before the trenches are *Testing* filled in to detect *drains.* any defectsin the pipes or joints. These should be made good and the test again applied until the whole system is in perfect order. Cement joints should be allowed to set for at least fortyeight hours before the test is made. There are several methods of testing. For the stoneware drains laid under the ground the *water test* is generally adopted. After the lower end of the length of drain to be tested has been securely stopped the drain is filled with water from its upper end until the desired pressure is obtained.

To obtain the required head of water extra lengths of pipe are sometimes taken up temporarily at the upper end of the drain or, as an alternative, both ends of the pipe may be plugged and water introduced under pressure by a force pump through a small aperture provided in the plug. The exact pressure may then be ascertained by a water pressure gauge. An escape of water through some defective portion of the drain is indicated by the subsidence of the level of the water in the upper part of the drain or by a diminution of the pressure shown by the gauge. Then the defect must be located and remedied and the drain re-tested until all weak points are eliminated. This process must be repeated in each section of the drainage system until the whole is found to be sound and tight. It is not necessary to test drains laid with ordinary socket joints made in cement with a greater pressure than is obtained with a 5 ft. or 6 ft. head of water. A foot head of water gives at its base a pressure of.433 lb per square inch, so that a head of 6 ft. would result in a pressure of just over lb per square inch. Cast-iron drain-pipes with caulked lead joints will withstand a pressure of nearly 90 lb per square inch of internal surface, but in actual practice it is sufficient if they are tested with a pressure of or say a head of 20 to 24 ft. The *atmospheric* or *air test* is sometimes applied instead of the water test. The drain is plugged, as in the latter, and air is then pumped into the pipes until the desired pressure is registered by the gauge attached to the apparatus. This pressure should be maintained without appreciable diminution for a stipulated period before the drains are passed as sound. The *smoke test* is generally used for testing vertical shafts such as soil-pipes and ventilators to which the water test cannot be conveniently applied owing to the excessive pressure produced at the lower portion of the pipe by the head of water. It is applied by stopping the ends of the pipes and introducing smoke by a drain rocket or by a smoke-producing machine which forces volumes of thick smoke through an aperture in the stopper. The pipes and joints are then carefully inspected for any evidence of leakage.

The *scent test* is occasionally employed for testing soil and ventilating pipes, but the apparatus must be carefully handled to avoid the material being spilt in the building and thus misleading the operator. The test is made by introducing into the drain some substance possessing a powerful odour such as oil of peppermint, calcium carbide or other suitable material, and tracing any defect by means of the escaping odour. This is not so effective a method as the smoke test, as there is more difficulty in locating leakages. Gulleys, traps and other similar fittings should be tested by pouring in water and observing whether siphonage or unsealing occurs. This of course will not happen if the appliances are of good design and properly ventilated. A section of a drain plug or stopper. It has a band of india-rubber which expands when the screw is turned and presses tightly against the inside of the drain-pipe. In the centre of the plug is a capped aperture which allows for smoke testing and also allows the water gradually to escape after a test by water.

Fig. Drain Stopper.

Existing drains which have become defective and require to be made good must be exposed, taken up and relaid with new pipes, unless advantage be taken of a method which, it is claimed, renders it possible to make them permanently watertight so as to withstand the water test under pressure, and at the same time to disinfect them and the surrounding subsoil. This end is accomplished with the aid of patent machines which on being passed through the drain-pipe first remove all obstructions and accumulations of foul matter and then thoroughly cleanse and disinfect it, saturating the outside concrete and contaminated soil adjacent to any leak with strong disinfectants. Subsequently, loaded with the best Portland cement, another machine is passed through the drain, and, by powerful evenly-distributed circular compression, forces the cement into every hole, crack or crevice in the pipes and joints.

This work leaves the inner surface of the pipes perfectly clean and smooth. After the usual time has been allowed for the cement to set the air test is applied, and the drain is claimed to be equal to, if not better than, a new drain, because the foundation is not dis turbed by the process, and the risk of settlement, which is often the cause of leaky drains, is remote. Every sanitary fitting should be trapped by a bend on the waste-pipe; this is generally made separately and fixed up near to the sink, closet or basin, as the case may be. The traps of small wastes such as those of sinks and lavatories should be fitted with a brass screw cap to facilitate clearing when a stoppage occurs. Their object is to hold a quantity of water sufficient to prevent the access of foul air through the waste-pipe into the house. The depth of the water " seal " should not be less than 2 in., or it may become easily unsealed in hot weather through the evaporation of the water.

Unsealing may be caused, too, by " siphonage," when a number of fittings are attached to the same main waste without the branches being properly ven tilated just below each trap. The discharge from one fitting in this case would create a partial vacuum in the other branches and probably suck the sealing water from one or more of the traps. To obviate such an occurrence an " anti-siphonage " pipe is fixed having its upper end open to the air and provided with branches tapping such waste-pipe just below the trap. Then, with this contrivance, a discharge from any fitting, instead of causing air to be sucked in through the trap of another fitting, thereby breaking the seal and allowing foul drain air to enter the house, merely draws the necessary air through the anti-siphonage pipe, leaving the other traps with their seals intact.

There are many forms of traps for use in different positions although the principle and purposes of all are identical. Two forms commonly used are known as the S and the P trap. The bell trap and the D trap are obsolete. To collect the rain and waste water from areas, yards, laundry and other floors and similar positions an open trapped gulley is used. It is usually of stoneware and fitted with an open iron grating which admits the water. Many of these gulleys are made too shallow and speedily get choked if the water they receive is charged with mud or sand. The gulleys are made with a deep container and are often fitted with a perforated basket of galvanized iron which catches the solid matter and has a handle which allows for its easy removal when necessary. Gulleys with slipper or channel heads are required to be fitted in some districts to receive the waste from sinks. The warm waste water from scullery and pantry sinks contains much grease, and should discharge into a trapped gulley specially constructed to prevent the passage of the grease into the drain. It should be of ample size to contain sufficient cold water to solidify the fat which enters it. This forms in cakes on the top of the water and should be frequently broken up and removed. Great attention has been directed to the design of sanitary fittings, with the object of making them as nearly self-cleansing as possible.

In the fixing of closets the wood casings which used to be fixed around every water-closet are going steadily out of use, their place being taken by a hinged seat supported on metal brackets - an arrangement which allows every part of the appliance 'to be readily cleaned with a cloth. In hospitals and similar institutions a form of closet is made fitted with lugs which are built into the wall; in this way support is obtained without any assistance from the floor, which is left quite clear for sweeping.

Lavatory basins and sinks are also supported on cantilevers in the same way, and the wood enclosures which were formerly often fixed around these appliances are now generally omitted. There are several distinct types of water-closets. Each type is made in many different patterns, both good and bad from a sanitary point of view, and, whatever the type decided upon, care is necessary in selecting to obtain one efficient and hygienic in shape and working. The principal kinds of closets now in use are the washdown, siphonic, valve, washout and hopper.

Washdown closets are most commonly used. They are inexpensive to buy and to fix, and being made in one piece and simple in construction without any mechanical work ing parts are not liable to get out of order. When strongly made or protected by brick or concrete work they will stand very rough usage. The objection is sometimes raised with regard to washdown closets that they are noisy in action. Washdown. designs have been greatly improved in this respect, and when fitted with a silent flushing cistern are not open to this objection.

Siphonic closets are a type of washdown in which the contents of the pan are removed by siphonic action, an after flush arrangement providing for the resealing of the trap. They are practically silent in action and with a flush of three gallons work very satisfactorily. Where the restrictions of the water company require the usual two gallon flush the ordinary washdown pan should be used.

Valve closets are considered by many authorities on sanitation to be preferable to all other types. For domestic buildings, hotels; and where not subjected to the hardest wear, they are undoubtedly of great value. They should have a three gallon flush, and on this account they cannot be used in many districts owing to the water companies' regulations stipulating that a flush of not more than two gallons may be used.

The *washout closet* is a type that never attained much popularity as it has been found by practical experience to be unsanitary and objectionable. The standing water is too shallow, and the receiving basin checks the force of the flush and the trap is therefore frequently imperfectly cleared.

ORIGINS OF SEWAGE

Sewage is generated by residential, institutional, and commercial and industrial establishments. It includes household waste liquid from toilets, baths, showers, kitchens, sinks and so forth that is disposed of via sewers. In many areas, sewage also includes liquid waste from industry and commerce. The separation and draining of household waste into greywater and blackwater is becoming more common in the developed world, with greywater being permitted to be used for watering plants or recycled for flushing toilets.

Sewage may include stormwater runoff. Sewerage systems capable of handling stormwater are known as combined sewer systems. This design was common when urban sewerage systems were first developed, in the late 19th and early 20th centuries. Combined sewers require much larger and more expensive treatment facilities than sanitary sewers. Heavy volumes of storm runoff may overwhelm the sewage treatment system, causing a spill or overflow. Sanitary sewers are typically much smaller than combined sewers, and they are not designed to transport stormwater. Backups of raw sewage can occur if excessive infiltration/inflow (dilution by stormwater and/or groundwater) is allowed into a sanitary sewer system. Communities that have urbanized in the mid-20th century or later generally have built separate systems for sewage (sanitary sewers) and stormwater, because precipitation causes widely varying flows, reducing sewage treatment plant efficiency.

As rainfall travels over roofs and the ground, it may pick up various contaminants including soil particles and other sediment, heavy metals, organic compounds, animal waste, and oil and grease. Some jurisdictions require stormwater to receive some level of treatment before being discharged directly into waterways. Examples of treatment processes used for stormwater include retention basins, wetlands, buried vaults with various kinds of media filters, and vortex separators (to remove coarse solids).

REMOVAL OF LARGE OBJECTS FROM INFLUENT SEWAGE

In primary treatment, the influent sewage water is strained to remove all large objects that are deposited in the sewer system, such as rags, sticks, tampons, cans, fruit, etc. This is most commonly done with a manual or automated mechanically raked screen. The raking action of a mechanical bar screen is typically paced according to the accumulation on the bar screens and/or flow rate. The bar screen is used because large solids can damage or clog the equipment used later in the sewage treatment plant. The solids are collected in a dumpster and later disposed in a landfill.

Primary treatment also typically includes a sand or grit channel or chamber where the velocity of the incoming wastewater is carefully controlled to allow sand grit and stones to settle, while keeping the majority of the suspended organic material in the water column. This equipment is called a detritor or sand catcher. Sand, grit, and stones need to be removed early in the process to avoid damage to pumps and other equipment in the remaining treatment stages. Sometimes there is a sand washer (grit classifier) followed by a conveyor that transports the sand to a container for disposal. The contents from the sand catcher may be fed into the incinerator in a sludge processing plant, but in many cases, the sand and grit is sent to a landfill.

SEDIMENTATION

Many plants have a sedimentation stage where the sewage is allowed to pass slowly through large tanks, commonly called "primary clarifiers" or "primary sedimentation tanks". The tanks are large enough that sludge can settle and floating material such as grease and oils can rise to the surface and be skimmed off.

The main purpose of the primary clarification stage is to produce both a generally homogeneous liquid capable of being treated biologically and a sludge that can be separately treated or processed. Primary settling tanks are usually equipped with mechanically driven scrapers that continually drive the collected sludge towards a hopper in the base of the tank from where it can be pumped to further sludge treatment stages.

SECONDARY TREATMENT

Secondary treatment is designed to substantially degrade the biological content of the sewage such as are derived from human waste, food waste, soaps and detergent. The majority of municipal and industrial plants treat the settled sewage liquor using aerobic biological processes. For this to be effective, the biota requires both oxygen and a substrate on which to live. There are number of ways in which this is done.

In all these methods, the bacteria and protozoa consume biodegradable soluble organic contaminants (*e.g.* sugars, fats, organic short-chain carbon molecules, etc.) and bind much of the less soluble fractions into floc. Secondary treatment systems are classified as fixed film or suspended growth. Fixed-film treatment process including trickling filter and rotating biological contactors where the biomass grows on media and the sewage passes over its surface. In suspended growth systems—such as activated sludge—the biomass is well mixed with the sewage and can be operated in a smaller space than fixed-film systems that treat the same amount of water. However, fixed-film systems are more able to cope with drastic changes in the amount of biological material and can provide higher removal rates for organic material and suspended solids than suspended growth systems.

Roughing filters are intended to treat particularly strong or variable organic loads, typically industrial, to allow them to then be treated by conventional secondary treatment processes. Characteristics include typically tall, circular filters filled with open synthetic filter media to which wastewater is applied at a relatively high rate. They are designed to allow high hydraulic loading and a high flow-through of air. On larger installations, air is forced through the media using blowers. The resultant wastewater is usually within the normal range for conventional treatment processes.

ACTIVATED SLUDGE

In general, activated sludge plants encompass a variety of mechanisms and processes that use dissolved oxygen to promote the growth of biological floc that substantially removes organic material. The process traps particulate material and can, under ideal conditions, convert ammonia to nitrite and nitrate and ultimately to nitrogen gas,.

Surface-aerated Basins

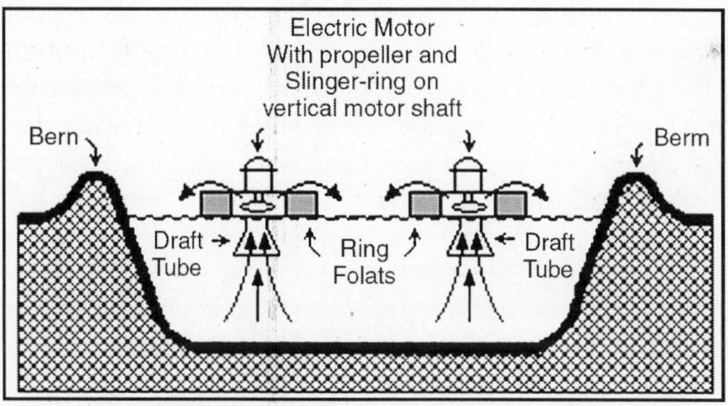

Fig. Typical Surface of Aerated Basin

Most biological oxidation processes for treating industrial wastewaters have in common the use of oxygen (or air) and microbial action. Surface-aerated basins achieve 80 to 90 per cent removal of Biochemical Oxygen Demand with retention times of 1 to 10 days.

In an aerated basin system, the aerators provide two functions: they transfer air into the basins required by the biological oxidation reactions, and they provide the mixing required for dispersing the air and for contacting the reactants (that is, oxygen, wastewater and microbes). Typically, the floating surface aerators are rated to deliver the amount of air equivalent to 1.8 to 2.7 kg. Biological oxidation processes are sensitive to temperature and, between 0 °C and 40 °C, the rate of biological reactions increase with temperature. Most surface aerated vessels operate at between 4 °C and 32 °C.

FILTER BEDS (OXIDISING BEDS)

In older plants and plants receiving more variable loads, trickling filter beds are used where the settled sewage liquor is spread onto the surface of a deep bed made up of coke (carbonised coal), limestone chips or specially fabricated plastic media. Such media must have high surface areas to support the biofilms that form. The liquor is distributed through perforated rotating arms radiating from a central pivot. The distributed liquor trickles through this bed and is collected in drains at the base. These drains also provide a source of air which percolates up through the bed, keeping it aerobic.

Biological films of bacteria, protozoa and fungi form on the media's surfaces and eat or otherwise reduce the organic content. This biofilm is grazed by insect larvae and worms which help maintain an optimal thickness. Overloading of beds increases the thickness of the film leading to clogging of the filter media and ponding on the surface.

CONVEYANCE OF SEWAGE

For small sewers, circular pipes of glazed stoneware or of moulded cement are used, from 6 in. to 18 in. and even in. in diameter. The pipes are made in short lengths, and are usually jointed by passing the end or spigot of one into the socket or faucet of the next. Into the space between the spigot and faucet a ring of gasket or tarred hemp should be forced, and the rest of the space filled up with cement. Other methods of jointing have already been described and illustrated. The pipes are laid with the spigot ends pointing in the direction of the flow, with a uniform gradient, and, where practicable, in straight lines. In special positions, as under the bed of a stream, castiron pipes are used for the conveyance of sewage.

Where the capacity of an 18-in. circular pipe would be insufficient, built sewers are used in place of stoneware pipes. These are sometimes circular or oval, but more commonly of an egg-shaped section, the invert or lower side of the sewer being a curve of shorter radius than the arch or upper side. The advantage of this form lies in the fact that great variations in the volume of flow must be expected, and the eggsection presents for the small or dry-weather flow a narrower channel than would be presented by a circular sewer of the same total capacity. Built sewers are most commonly made of bricks, moulded to suit the curved structure of which they are to form part. Separate invert blocks of glazed earthenware, terra-cotta or fire-clay are often used in combination with brickwork. The bricks are laid over a templet made to the section of the sewer, and are grouted with cement. The thickness of brickwork for sewers over 3 ft. in diameter should not be less than q in., but for smaller sewers laid in good ground at depths not exceeding 20 ft. from the surface a thickness of 41 in. will suffice if well backed up with concrete. The thickness of brickwork for a sewer of any size may be determined in feet by the formula dr/ioo, where d= depth of excavation in feet and r = external radius in feet.

An egg-shaped sewer, made with two thicknesses of brick, an invert block, and a concrete setting. Concrete is largely used in the construction of sewers, either in combination with brickwork or alone. For this purpose the concrete consists of from 5 to 7 parts of sand and gravel or broken stone to i of Portland cement. It may be used as a cradle for or as a backing to a brick ring, or as the sole material of construction by running it into position round a mould which is removed when the concrete is sufficiently set, the inner surface of the sewer being in this case coated with a thin layer of cement.

A developthe construction of concrete sewers, whether laid in pipes or constructed and moulded *in situ,* is the use of iron or steel bars and wires embedded in the material as a reinforcement. Such conduits can be constructed of any size and designed to withstand high pressures a section of a concrete sewer having a diameter of more than 9 ft. constructed with round rod reinforcement. With regard to the method for calculating the proportions, generally speaking the thickness of the concrete shell should in no place be less than one-twelfth of the greatest internal diameter of the tube, while the steel reinforcement should be designed to resist the whole of the tensile stress. Where the safe tensile stress in the steel is 8 tons per sq. in. P= the pressure in pounds per sq. in., and *r*= the internal radius in inches; the weight of the reinforcement per sq. ft = Pr/450, while its area at each side of the pipe per longitudinal foot, when *f*= *safe* tensile stress in the reinforcement in pounds, is I 2 Pr/f.

In determining the dimensions of sewers, the amount of sewage proper may be taken as equal to the water supply (generally about 30 gallons per head per diem), and to this must be added *D* (when the " combined " system is adopted) an allowance for the surface water due to rainfall. The latter, which is generally by far the larger constituent, is to be estimated from the maximum rate of rainfall for the district and from the area and character of the surface. In the sewerage of Berlin, for example, the maximum rainfall allowed for is $ of an inch per hour, of which one-third is supposed to enter the sewers. In any estimate of the size of sewers based on rainfall account must of course be taken of the relief provided by storm-overflows and also of the capacity of the sewers to become simply charged with water during the short time to which very heavy showers are invariably limited.

Rainfall at the rate of 5 or 6 in. per hour has been known to occur for a few minutes, but it is unnecessary to provide (even above storm-overflows) sewers capable of discharging any such amount as this; the time taken by sewers of more moderate size to fill would of itself prevent the discharge from them from reaching a condition of steady flow; and, apart from this, the risk of damage by such an exceptional fall would not warrant so great an initial expenditure. Engineers differ widely in their estimates of the allowance to be made for the discharge of surface water, and no rule can be laid down which would be of general application.

In order that sewers should be self-cleansing, the mean velocity of flow should be not less than 22 ft. per second. The gradient necessary to secure this is calculated on principles which are stated in the article Hydraulics. The velocity of flow, V, is V =*G-Vim,* where *i* is the inclination, or ratio of vertical to horizontal distance; m is the " hydraulic mean depth," or the ratio of area of section of the stream to the wetted perimeter; and *c* is a coefficient depending on the dimensions and the roughness of the channel and the depth of the stream. A table of values of *c* will be found in § 98 of the article referred to. This velocity multiplied by the area of the stream gives the rate of discharge. Tables to facilitate the determination of velocity and discharge in sewers of various dimensions, forms and gradients will be found in Latham's and other practical treatises.

Where the contour of the ground does not admit of a sufficient gradient from the gathering ground to the place of destination, the sewage must be pumped to a higher level at one or more points in its course. To minimize this necessity, and also for other reasons, it is frequently desirable not to gather sewage from the whole area into a single main, but to collect the sewage of higher portions of the town by a separate highlevel or interception sewer. It is undoubtedly necessary to construct overflows for storm water in connexion with combined systems of sewerage.

No combined sewer of such size as will make it comparatively Storm-self-cleansing under normal conditions can hope to carry water off the volume of water resulting from heavy rain. It might be thought that the overflow resulting from a storm would consist of nearly pure rain-water, but this is not the case, as the pressure of storm water has the effect of scouring out from the sewers a great deal of foul matter that is deposited when the flow is small. This being the case it is obviously bad policy to take the overflow into a stream, which would thereby suffer contamination.

A better plan is to direct the discharge into a dry ditch or channel where the liquid may soak into the soil and the solid particles by contact with the air may quickly become oxidized. In agricultural districts it might be possible by arrangement with farmers to run the overflow over grass-land, as it has good manurial properties. Occasionally when a sewer has to cross a stream or other obstruction it is found impossible to bridge or carry the pipe across and preserve its proper gradient. In such cases it must be carried under the obstruction by means of an inverted *s,* siphon. The exact form that should be given to inverted *p* siphons is disputed, but it is generally agreed that they are expedients to be avoided wherever possible. The majority take roughly the form of the stream section, that is, they have two sloping pieces corresponding with the banks with a flat cross-piece under the bed of the stream. The pipes are invariably of iron and should be laid in duplicate, as they are liable to silt up in the flat length.

For this reason it is usual in constructing a siphon to place permanent chains in the pipes, and these are periodically pulled backward and forward to stir up the silt. Brushes may also be attached to the chains and pulled through from end to end. At either end of the siphon pipes there are manholes into which the pipes are built. Penstock valves also should be provided at each end so that sewage can be shut out of one or both of the siphons as desired for clearing purposes.

Tumbling bays being prohibited, the usual method of leading a high-level sewer into a low-level sewer is by means of a ramp. This is constructed in connexion with a manhole into which Con- the end of the high-level sewer is taken and finished ' onio usually with a flap valve. Some distance back along th this sewer a wide-throated junction is put in the invert of - the sewer, and from this junction a ramp-pipe is taken down to the invert of the low-level sewer, so that the sewage in the upper sewer instead of having a direct fall runs down the slope of the ramp. The ramp-pipe is usually constructed of iron and is of smaller section than the high-level sewer because of the greater fall and pressure.

In the low-lying parts of towns storage tanks are often constructed to receive the sewage of such districts. They are periodically emptied of their contents, which are pumped up into the main sewers through which the sewage travels to the outfall. This storing of sewage should be avoided whenever possible. It is much better to provide for raising it as it is produced either by an installation of one or more automatic lifts, such as Adams's sewage lifts, or, where a large amount of material is to be dealt with, necessitating continual pumping, by a Shone ejector worked by compressed air.

Sewer gas is a term applied to the air, fouled by mixture with gases which are formed by the decomposition of sewage, and by the organic germs which it carries in suspension, that fills the sewer in the variable space above the liquid stream. It is universally recognized that sewer gas is a medium for the conveyance of disease, and in all well-designed sewers. systems of sewerage stringent precautions are taken to keep it out of houses. It is equally certain that the dangerous character of sewer gas is reduced, if not entirely removed, by free admixture with the oxygen of fresh air. Sewers should be liberally ventilated.

Only for this reason, but to prevent the air within them from ever having its pressure raised (by sudden influx of water) so considerably as to force the " traps " which separate it from the atmosphere of dwellings. The plan of ventilation now most approved is the very simple one of making openings from the sewer to the surface of the street at short distances - generally shafts built of brick and cement - and covering these with metallic gratings. Under each grating it is usual to hang a box or tray to catch any stones or dirt that may fall through from the street, but the passage of air to and from the sewer is left as free as possible.

Fig. Brick Sewer.

The openings to the street are frequently made large enough to allow a man to go down to examine or clean the sewers, and are then called " manholes." Smaller openings, large enough to allow a lamp to be lowered for purposes of inspection, are called " lampholes," and are often built up of vertical lengths of drain-pipe, 6 in. or 9 in. in diameter, and finished at the surface with a cover similar to that used for a manhole but smaller. A length of 150 ft. of pipe sewer is about the limit that can be sighted through. Lampholes are mostly used in the construction of pipe and other small sewers. To facilitate inspection and cleaning, sewers are, as far as possible, laid in straight lines of uniform gradient, with a manhole or lamphole *Flashing* at each change of direction or of slope and at each junction *F sewers.* of mains with one another or with branches. The sewers may advantageously be stepped here and there at manholes.

Sir R. Rawlinson pointed out that a difference of level between the entrance and exit pipes tends to prevent continuous flow of sewer gas towards the higher parts of the system, and makes the ventilation of each section more independent and thorough. When the gradient is slight, and the dry-weather flow very small, occasional flushing must be resorted to. Flap valves or sliding penstocks are introduced at manholes; by closing these for a short time sewage (or clean water introduced for the purpose) is dammed up behind the valve either in higher parts of the sewer or in a special flushing chamber, and is then allowed to advance with a rush. Many self-acting arrangements for flushing have been devised which act by allowing a continuous stream of comparatively small volume to accumulate in a tank that discharges itself suddenly when full. A valuable contrivance of this kind is Rogers Field's siphon flush tank.

When the liquid in the tank accumulates so that it reaches the top of the annular siphon, and begins to flow over the lip, it carries with it enough air to produce a partial vacuum in the tube. The siphon then bursts into action, and a rapid discharge takes place, which continues till the water-lèvel sinks to the foot of the bellshaped cover. Adams's " Monster Flusher " is constructed on similar principles and is of simple and strong design. Its flushingpower is claimed to be greater than that of the ordinary siphon. By the use of this appliance, which is automatic in action, shallow sewers can be effectively flushed.

BIOLOGICAL AERATED FILTERS

Biological Aerated (or Anoxic) Filter (BAF) or Biofilters combine filtration with biological carbon reduction, nitrification or denitrification. BAF usually includes a reactor filled with a filter media. The media is either in suspension or supported by a gravel layer at the foot of the filter. The dual purpose of this media is to support highly active biomass that is attached to it and to filter suspended solids. Carbon reduction and ammonia conversion occurs in aerobic mode and sometime achieved in a single reactor while nitrate conversion occurs in anoxic mode. BAF is operated either in upflow or downflow configuration depending on design specified by manufacturer.

FLUIDIZED BED REACTORS

The carbon absorption following biological treatment is particularly effective in reducing both the BOD and COD to low levels. A fluidized bed reactor is a combination of the most common stirred tank packed bed, continuous flow reactors. It is very important to chemical engineering because of its excellent heat and mass transfer characteristics.

In a fluidized bed reactor, the substrate is passed upward through the immobilized enzyme bed at a high velocity to lift the particles. However the velocity must not be so high that the enzymes are swept away from the reactor entirely. This causes low mixing; these type of reactors are highly suitable for the exothermic reactions. It is most often applied in immobilized enzyme catalysis

MEMBRANE BIOREACTORS

Membrane bioreactors (MBR) combines activated sludge treatment with a membrane liquid-solid separation process. The membrane component uses low pressure microfiltration or ultra filtration membranes and eliminates the need for clarification and tertiary filtration. The membranes are typically immersed in the aeration tank (however, some applications utilize a separate membrane tank). One of the key benefits of a membrane bioreactor system is that it effectively overcomes the limitations associated with poor settling of sludge in conventional activated sludge (CAS) processes. The technology permits bioreactor operation with considerably higher mixed liquor suspended solids (MLSS) concentration than CAS systems, which are limited by sludge settling. The process is typically operated at MLSS in the range of 8,000–12,000 mg/L, while CAS are operated in the range of 2,000–3,000 mg/L. The elevated biomass concentration in the membrane bioreactor process allows for very effective removal of both soluble and particulate biodegradable materials at higher loading rates. Thus increased Sludge Retention Times (SRTs)—usually exceeding 15 days—ensure complete nitrification even in extremely cold weather.

The cost of building and operating a MBR is usually higher than conventional wastewater treatment, however, as the technology has become increasingly popular and has gained wider acceptance throughout the industry, the life-cycle costs have been steadily decreasing. As well, in developed urban areas where the footprint of the treatment plant is considered a limiting factor MBR facilities can be considered a desirable option.

SECONDARY SEDIMENTATION

The final step in the secondary treatment stage is to settle out the biological floc or filter material and produce sewage water containing very low levels of organic material and suspended matter.

Rotating Biological Contactors

Schematic diagram of a typical rotating biological contactor (RBC). The treated effluent clarifier/settler is not included in the diagram. Rotating biological contactors (RBCs) are mechanical secondary treatment systems, which are robust and capable of withstanding surges in organic load. RBCs were first installed in Germany in 1960 and have since been developed and refined into a reliable operating unit. The rotating disks support the growth of bacteria and micro-organisms present in the sewage, which breakdown and stabilise organic pollutants.

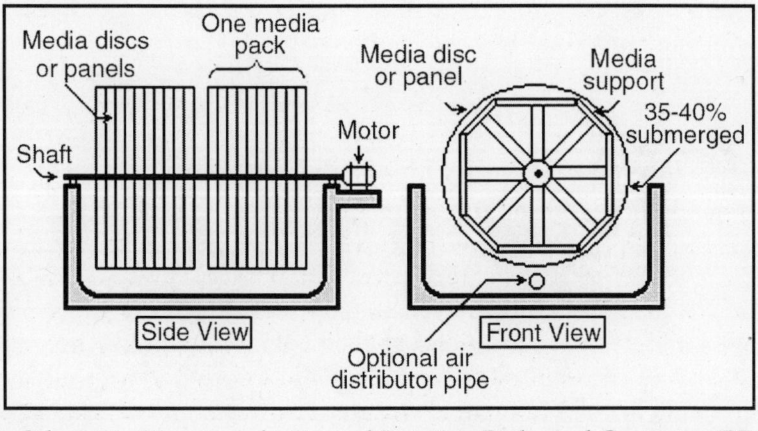

Fig. Schematic Diagram of a Typical Rotating Biological Contactor (RBC).

To be successful, micro-organisms need both oxygen to live and food to grow. Oxygen is obtained from the atmosphere as the disks rotate. As the micro-organisms grow, they build up on the media until they are sloughed off due to shear forces provided by the rotating discs in the sewage. Effluent from the RBC is then passed through final clarifiers where the micro-organisms in suspension settle as a sludge. The sludge is withdrawn from the clarifier for further treatment.

COMPOSITION DISPOSAL OF DOMESTIC SEWAGE

The composition of domestic sewage is now fairly well known and is generally reduced for the purposes of comparison to a standard; that is to say, ordinary sewage is that due to a water-supply of about 30 gallons per head per diem. If the supply is less, and there is no leakage of subsoil water into the drainage system, the sewage will be stronger; conversely, if there is leakage, &c., the sewage will be more dilute, but obviously, the quantity of impurities will, for any given population, be the same in amount. The subjoined table shows the kind of sewage referred to: A *verage Domestic Sewage, in Grains per Gallon.*

For all practical purposes we may say that average sewage contains two tons of suspended matters in each million gallons, one-half of which is mineral matter. When, however, we come to a consideration of trade waste, the question becomes difficult in the extreme, because of the great variety of trades, and the ever varying quantities added to the sewage. Some of the principal trade wastes are from dye-works, print-works, bleach-works, chemical works, tanneries, breweries, paper-makers, woollenworks, silk-works, iron-works and many others. In some cases one only of these trade wastes finds its way to the sewers; in others, several of them may be found. In some instances, again, these trade wastes are of an alkaline nature, in others they are acid; the mixtures may be either, and of greatly varying character. Next comes the manner in which sewage is discharged at the works.

The flow is variable throughout the entire 24 hours, but in the case of sewers discharging domestic sewage only, such sewage being of the standard strength, it will be a close approximation to the facts to say that about two-thirds is discharged between the hours of 7 A.M. and 7 P.M., one-half during the eight hours of maximum flow, two-fifths during the six hours of maximum flow, and about 71 per cent per hour during the two hours of maximum flow. These data will be sufficient for the design of the works intended for dealing with the sewage. Separate calculations must be made if there is trade refuse, or much leakage of subsoil water. In very large systems, again, the maxima are rather less because of the time occupied by the sewage in travelling to the outfall from the more remote parts of the district. In cases where one set of sewers is employed for both sewage and rainfall the sewage flow may be increased more than a hundredfold within a few minutes by heavy rainstorms.

Of course the sewage disposal works can only deal with a small proportion of such flow, and the balance is discharged into some convenient water-course or other suitable place. Even when the separate system is employed, as in the case of the smaller towns, the flow may be increased ten to fifteen times by rain, because it is unusual to carry two sets of drains to the backs of the houses. In designing outfall works, therefore, all these circumstances must be carefully considered. Again, when the sewage is pumped, as is frequently the case, the size of the tanks must often be increased, because in the smaller installations the whole of the day's sewage is frequently pumped out in a few hours; this fact must also be remembered when designing filters.

Nearly every town upon the coast turns its sewage into the sea. That the sea has a purifying effect is obvious. The object to be attained is its dispersion in a large volume of sea-water. As it is lighter than salt water it tends to rise after leaving the sewer; the outfall should, therefore, if practicable, terminate in deep water, so that the two liquids may become well mixed. The currents must be studied by means of floats, and in most cases the sewage must be discharged upon the ebb tide only, and then perhaps not throughout the entire period, the object being to prevent it from being carried towards the shore.

That the purification is effected mainly by means of living organisms is well established, and it has been urged by competent authorities that this system is not wasteful, since the organic matter forms the food of the lower organisms, which in turn are devoured by fish. Thus the sea is richer, if the land is the poorer, by the adoption of this cleanly method of disposal. The next step is the partial purification of the sewage by means of a chemical process. When a town lies some distance up an estuary, as for example London, Glasgow, Rochester and many others, the dilution may be insufficient to prevent a nuisance, or the suspended matters may be deposited upon the foreshore to be uncovered at low water. The first stage of purification is then employed, namely, clarification in tanks.

Practice varies with regard to tank capacity, but as a general rule it should be at least equal to half a day's dry weather flow. This will enable the works manager to turn out a good effluent, even in wet weather, when the volume is much increased. With regard to the practical effect of any particular treatment, it is now recognized that the matters in solution are scarcely touched by any chemical process that can be employed, but the removal of the suspended matter is a great manholes n cav per cent gain, as has been proved in the case of London. Briefly, a good chemical process will do about one-half of the work of purification; and in many cases it is not necessary to go further.

With regard to the kind of chemical to use, lime, either alone or in conjunction with aluminium sulphate or with ferrous sulphate, is most frequently employed. When the resulting sewage sludge has to be filter-pressed, lime is almost essential for the primary treatment of the sewage, in order to destroy the glutinous nature of the sludge. In the case of large towns like London, Manchester and Salford, the sludge is shipped in specially designed steamers, of 600 tons to r000 tons burden, and discharged into the sea at a distance from the coast. The London outfall works have a fleet of six steamers, which convey the sludge out to Barrow Deep, a channel in the North Sea about 10 m. east of the Nore lightship.

Each vessel has four oblong tanks having a total capacity of ‘coo tons of sludge, which can be discharged in seven minutes when the valves are fully opened. The sludge is discharged about io ft. under the water and being agitated by the action of the ship’s screws is very completely diffused. The sand and earthy matters soon subside and the organic matter is rapidly consumed by the organic life in the sea-water. A careful microscopical examination and chemical analysis failed to detect more than the merest trace of the mineral portion of the sludge, either in dredgings from the bottom of the channels or on the surface of the sandbanks. The cost of the disposal works out at about 41d. per ton of sludge.

In the case of towns situated on rivers above the range of tidal waters, the further purification is effected either on land, or by means of artificial filters, or a combination of the two. The question of land treatment is frequently considered from the standpoint of so many persons to the acre; but the best method is to ascertain how many gallons per day an acre of land will purify. As the quality of land varies greatly, the proper volume to be applied per acre can only be ascertained after a good deal of experience. The range lies between about 3000 gallons per acre per day in the case of poor land, to about 30,000 gallons in the same period in the case of the best. Let us assume an instance of the latter kind.

The works have been designed on a basis of 1000 persons per acre, producing 30,000 gallons of sewage per day; the land being of a highly suitable character, and the sewage having been clarified, success is assured. But, conversely, through faulty construction of the sewers, the sewage amounts, say, to 60 gallons per head; the land, unable to deal with the liquid, quickly becomes water-logged and offensive, and the works are a failure. Precisely the same remarks apply to artificial filters, which are always designed upon the basis of so many gallons per square yard of filtering material.

Many failures of both land and filters have been due to the fact that the actual sewage flow was greatly in excess of the original estimates. We may say that clay soils lie at one end of the scale, and very porous sands or gravels at the other; obviously, therefore, each case must be considered on its merits. It should be remembered that when such moderate quantities as 3000 gallons per acre per day are applied to land, there is no necessity to remove the suspended matter; broad irrigation being resorted to, the land readily assimilates the solids, and thus one source of expense may be eliminated.

The artificial filters are now generally called bacteria beds; although filters have been in constant use in some cases, as for instance at Wimbledon, for a great number of years. The first filters constructed at these works were made in 1876, and were about 7000 sq. yds. in extent. With the growth of population additions have been made of at least five times that area. One of the original beds was used for crude sewage, but the mineral matter choked it completely, and experience pointed to the necessity of clarifying the sewage before filtration.

Whether the treatment should be in open or in closed tanks, or whether chemicals should be added, has been much debated; but seeing that ordinary sewage contains one ton of suspended mineral matter in each million gallons, it is clear that if this is not removed before filtration, it will be retained in the filters and ultimately choke them, as happened at Wimbledon. The common cesspool has been resuscitated and improved under the name of a septic tank. In this the disintegration of the suspended matter is brought about by anaerobic organisms, and the liquid in passing slowly through the tank absorbs most of the gases due to the breaking down of the organic matter. There is no oxidation at this stage. The liquid is next passed through artificial filters, of which there are many types. What is known as a " contact " filter was constructed, probably for the first time on a large scale, at the London (Barking) works.

The object sought to be attained was that of making each cubic yard of filtering material perform the same amount of work, and the least expensive way was apparently to close the outlet, and charge the filter with liquid, allowing it to remain in contact for about two hours, and then drawing it off so that the bed could be thoroughly aerated. No doubt a better way would be to distribute the sewage in the form of a shower of liquid, and work the beds continuously, but this involves a good deal of expense for spreading appliances, and a fall is necessary in the works, which is not always obtainable. Probably the most complete installation of the kind last referred to is that at Salford.

Iron pipes are led over the surface of the filters, and spraying nozzles are placed at short intervals, so that the sewage is applied in the form of a heavy shower. But whatever form the filters and appliances may assume, the final result is the same. If the beds are properly aerated, the aerobic organism establishes itself in prodigious numbers, and attacks the organic matter, breaking it down into harmless, soluble and gaseous products. It is, of course, assumed that the filters are adequate in area, and are properly managed. With regard to the materials to be employed in making sewage filters, it is now well established that the size of the particles has a more important bearing than their composition. At the same time, it may be remarked that materials with very rough surfaces, as for instance coke breeze, are more effective than those with smooth surfaces. Doubtless the former classes afford, in the interstices, a lodging for the bacteria, and no doubt a given quantity of material with rough surfaces will harbour greater numbers than the same amount of smooth. A reference must be made to the Manchester experiments. The experts' report suggested the provision of 60 acres of filters for dealing with the sewage of the city, which is said to average 30 million gallons per day in dry weather. But after inquiry into the merits of the proposal the officials of the Local Government Board recommended that the filters should be 92 acres in extent, and that the effluent should be finished on land.

WASTEWATER AND STORMWATER COLLECTION IN SEWERAGE SYSTEM

A sewerage system collects wastewater and can be in the form of blackwater separated from greywater, or mixed with it (sewage). Gravity is used wherever possible to convey the wastewater. It is not surprising therefore that natural stormwater drainage is usually used, because this is how rainwater run-off is conveyed in nature by gravity.

The principle of using gravity as the driving force for conveying wastewater in a sewerage system should be applied wherever possible, because this will minimise the cost of pumping. Natural stormwater drainage occurs in what is usually termed a catchment basin. In a catchment basin, rainwater run-off flows to a common point of discharge, and in so doing, forms streams and rivers. Crossing a catchment boundary may mean that the water has to be unnecessarily pumped, requiring an energy source. A wastewater sewerage system should therefore be within a stormwater catchment basin. Sewerage systems can be classified into combined sewerage and separate sewerage.

Combined sewerage carries both stormwater and wastewater, while separate sewerage carries stormwater or wastewater separately. Recent trends have been for the development of separate sewerage systems. The main reason for this is that stormwater is generally less polluted than wastewater, and that treatment of combined wastewater and stormwater is difficult during heavy rainfalls, resulting in untreated overflows (commonly termed combined sewer overflow, CSO). In practice there is usually ingress of stormwater into wastewater sewerage pipes, because of unsealed pipe joints, and unintentional or illegal connections of rainwater run-off. Conversely there may be unintentional or illegal wastewater connections to stormwater sewerage.

Wastewater sewerage systems can be classified into three major types:

1. Conventional sewerage
2. Simplified sewerage
3. Settled sewerage

CONVENTIONAL SEWERAGE

Conventional sewerage is also termed deep sewerage because the sewerage pipes are laid deep beneath the ground. Pumping is generally required at various stages of the sewer pipe network, especially if the landscape is fairly flat.

The larger the population served by the sewerage system, and the longer the planning horizon is to cope with future population increases, the larger the diameter of the final pipes becomes. The costs of the pipes, inspection manholes, pumps and pumping stations and their construction/installation are therefore high. The costs of operation and maintenance are correspondingly high because of very conservative design assumptions.

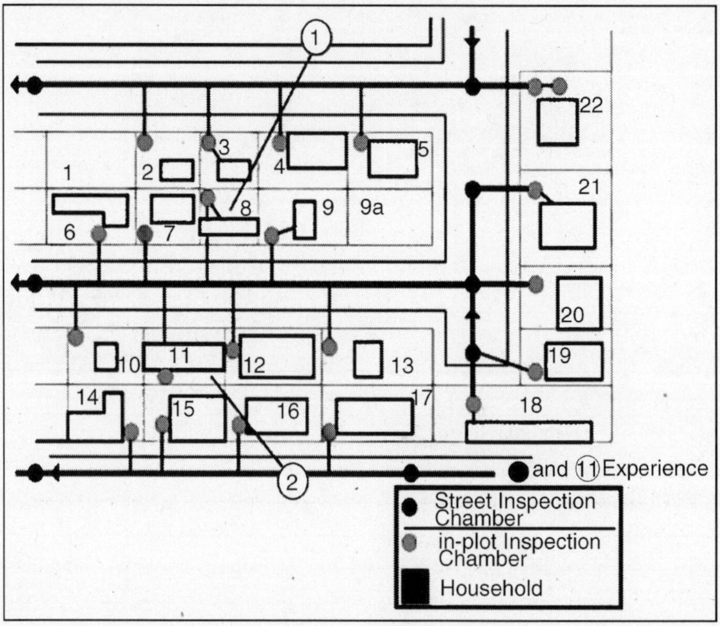

Fig. Pipe Layout for Conventional Sewerage

Simplified Sewerage

Simplified sewerage is also known as shallow sewerage, reflecting the shallower placement of the pipes in contrast to the conventional or deep sewerage. The purpose of simplified sewerage is to reduce the cost of construction and the corresponding cost of operation and maintenance.

Simplified sewerage design is based on hydraulic theory in the same manner as for conventional sewerage but has less conservative design assumptions. Smaller diameter pipes are used when water use per person is known to be less and the minimum depth of cover of pipes can be as low as 0.2 m when there is only light traffic. Manholes can be replaced by inspection cleanouts because of the shallow pipes. The design planning horizon can be shorter because the population projection may be uncertain. In a variation of the simplified sewerage, the pipe layout passes through property lots (condominial) rather than on both sides of a street (conventional). Figures show the sewerage layouts in conventional sewerage and in condominial sewerage. The cost of construction of simplified sewerage can be 30 to 50 per cent less than conventional sewerage depending on local conditions.

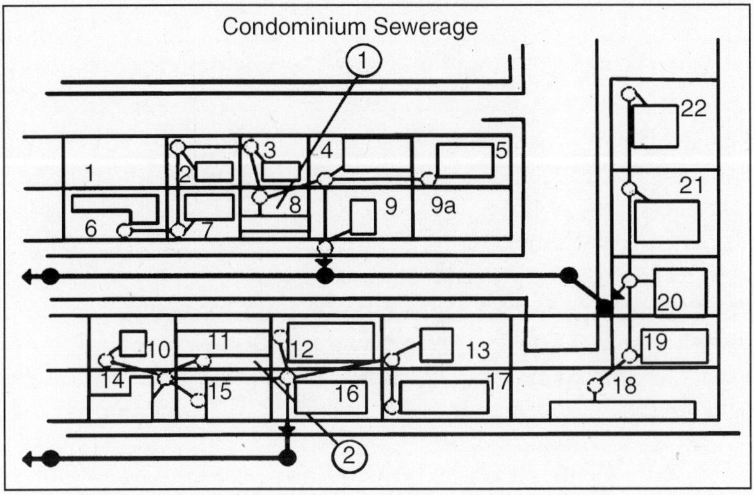

Fig. Pipe Layout for Condominimum Sewerage

Shallow sewerage is also conducive to local community participation because sewer pipes have to cross property boundaries. The community has to agree to this arrangement which extends after construction for maintenance. The shallow pipe, and hence the shallow trenches, also allow members of the community to participate by, for example, providing labour for digging the trenches.

This is in contrast to conventional sewerage where specialised machinery is required for the deep trenches. Simplified sewerage was originally developed in Brazil and is increasingly being used in other parts of the world. The 'International Source Book on Environmentally Sound Technologies for Wastewater and Stormwater Management', published by IWA and IETC, provides useful case studies.

Settled Sewerage

Settled sewerage refers to sewerage for conveying wastewater that has been settled, for example, in a septic tank. Settled sewerage originated to convey the overflow from septic tanks where the soil cannot cope or absorb the overflow. This usually occurs when the groundwater table is high, or where the soil permeability is low, or where there are rock outcrops. It can also be used when effluent from septic tanks pollutes groundwater and it is necessary to convey the effluent off-site and treat it. Because there are no solids that can potentially sediment in the sewerage pipes, there is no requirement for the self-cleansing velocity. Smaller pipes and lower gradients can be used. The cost of settled sewerage is between a third and a half of conventional sewerage. Originally developed in South Australia to overcome problems with failing septic tanks, it has been used quite widely worldwide to upgrade septic tank systems.

TREATMENT OF WASTEWATER AND STORMWATER

The treatment of wastewater and stormwater means the reduction and removal of pollutants from the water. The first principle to bear in mind therefore is to prevent pollutants from entering the water in the first place. In the case of stormwater we need to ensure that surfaces through which stormwater run-off passes over should as far as possible be free from solids and other wastes. Thus the collection of solid wastes is an important part of stormwater treatment as is the separate collection of wastewater and stormwater. The treatment of industrial wastewaters before discharge to the sewer is also extremely important in preventing pollutants from entering the wastewater and stormwater systems.

In the case of wastewater, separating blackwater and greywater can mean less energy is required in treatment. This is because blackwater contains most of the solids, which during treatment have to be removed from the mixture. Further separating urine and faecal materials may also mean that the urine can be reused without much treatment and the faecal materials can be more simply treated. The use of water to convey toilet wastes may be questioned based on this principle, because treatment means separating these wastes from the water. Besides preventing pollutants entering the water, water conservation means that a lower volume of water has to be treated. Since the size of treatment systems is primarily governed by the volume of water to be treated rather than the amount of pollutants in the water, a lower volume means smaller treatment plants and a corresponding capital cost. Use of less water to flush toilets belongs to this principle.

A range of wastewater treatment technology options is presented below. Treatment of wastes on-site is considered first followed by off-site treatment of the wastewater. Each technology requires maintenance and proper operation. The demand of each technology for maintenance varies and this is also discussed, as well as the public health and environmental impacts of the technology. Where there is no existing septic tank, an interceptor box or tank can be used. It functions like a septic tank and designed in the same way. To reduce cost, the wastewater from a group of houses can be connected to one interceptor tank. Just like in a septic tank, the accumulation of sludge has to be removed regularly from an interceptor tank.

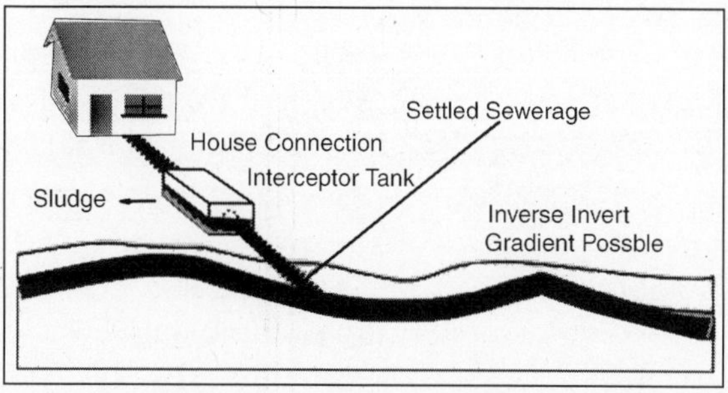

Fig. Interceptor Tank in Settled Sewerage

THE LANDSCAPE'S NATURAL DRAINAGE SYSTEM

Stormwater Collection

Stormwater flows through the landscape's natural drainage system. Piped stormwater collection was a development in European cities to overcome odour and improve aesthetic appearance of wastewater disposed with stormwater. The covering of ditches used for combined sewerage was an intermediate step in using natural drainage to construct sewerage for combined wastewater and stormwater. Piped sewerage also allows more land area for road and footpaths. With the separate collection of wastewater there is an opportunity to return some stormwater flow path to its more natural state to improve urban amenity value.

On-site Wastewater Treatment Systems

On-site treatment relies on decomposition of the organic wastes in human excreta by bacteria. This can take place in a simple pit in the ground or in specially designed tanks to promote the bacterial decomposition of the wastes. Unless re-use of the wastewater is specifically intended, the overflow from the pit or tank is allowed to soak into the ground. Further bacteriological decomposition and soil filtration, absorption and purification processes take place in the soil. The potential for groundwater pollution, however, exists with on-site treatment and disposal systems, because not all pollutants (*e.g.* nitrate) are removed by these processes. Pit latrine, pour flush latrine, composting toilet, septic tank and two improved on-site treatment units are described below because they represent major types of on-site treatment systems. Variations of these exist and are used in different areas of the world. Some of these are described in greater detail in the Regional Overviews in the Source Book published by IWA and IETC. The treatment principles are, however, covered under these major types.

Pit Latrine

A pit latrine collects excreta in a pit dug in the ground beneath the toilet structure. If the soil is loose the pit needs to be lined with, for example, loose bricks to prevent the wall from collapsing. During storage in the pit decomposition of the organic substances takes place under anaerobic conditions. The anaerobic decomposition releases gases (carbon dioxide, methane and sulphuric gases) and reduces the volume of sludge. Seepage of water into the surrounding soil takes place through the sides and bottom of the pit. During seepage further decomposition of organic matter by soil bacteria takes place reducing the BOD of the water. There will also be die-off of bacteria and viruses during storage and as the water percolates through the soil. Bacteria under these conditions do generally not remove nutrients, so pollution of groundwater will occur.

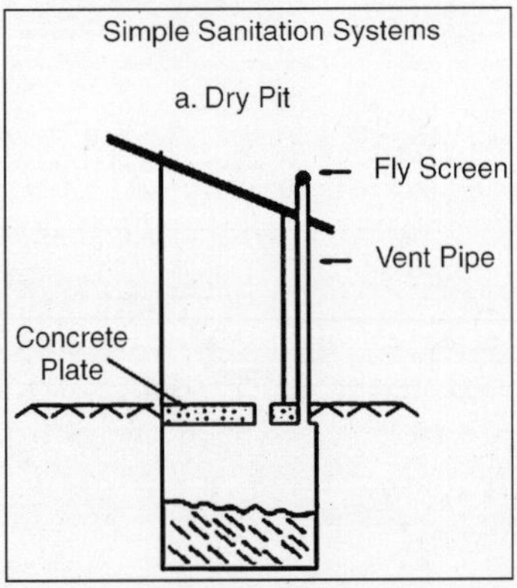

Control of odour and insects are important with a pit latrine. This is achieved by having a vented pit. The vent acts to draw odour and insects into the pit and up the vent. Gases (methane and carbon dioxide) produced by the decomposition of the excreta also leave through the vent. Ensuring that the vent protrudes well above the roof of the housing allows ventilation through natural convection. Facing the vent towards the sun (southward in the Northern hemisphere and northward in the Southern hemisphere) and painting the vent black to maximise absorption of heat from the sun will help venting by heat convection.

The heated air in the vent rises and draws air from the toilet. Ventilated improved pit (VIP) toilets are widely used in Africa. Pit latrines pose problems when groundwater is shallow and the pit is in groundwater or close to it. There is no soil barrier to protect the water quality of the groundwater, and mosquitoes may breed inside the pit. A pit is also difficult to dig when the ground is rocky. Pit latrines should not be used in these cases. The pit will eventually fill with faecal sludge and needs to be emptied. The period between emptying depends on the size of the pit and its usage. It is desirable to design the pit to store at least one year of sludge production. Emptying requires mechanical suction of the sludge. The sludge requires treatment prior to re-use or disposal. Two adjoining pits can be used alternately. Further decomposition of sludge in a full pit takes place while the adjacent pit is in use.

Its content after further decomposition can be manually removed. An alternative way of dealing with a full pit is to dig another pit and relocate the sanitary platform and toilet housing to the new pit. The full old pit can then be covered with soil, preferably of greater than 15 cm depth to prevent disease vectors (rodents and insects) from burrowing into it.

Composting Toilet

Rather than the decomposition of the faecal sludge under anaerobic conditions (no oxygen) in the pit of a pit latrine, decomposition under aerobic conditions (with oxygen) can be promoted in an above ground (elevated) latrine. Air can be introduced through an opening to pass through the sludge and exit through the vent, while excess liquid is allowed to drain for collection or evaporation. With two adjoining composting chambers or vaults used alternately, the process of composting in an already full chamber can be allowed to proceed until the chamber is to be used again, and produce mature compost for direct re-use in the garden.

Other household organic wastes (*e.g.* food wastes) can be added to the faecal sludge, and materials such as newspaper or sawdust can be added to balance the carbon to nitrogen ratio for optimal composting. Because mature compost takes several months to produce under ambient temperatures, it is desirable for the chambers to be sized to hold at least 6 months of waste. Worms can also be added to assist with vermi-composting. Further details on handling and composting of sludge can be found in Section.

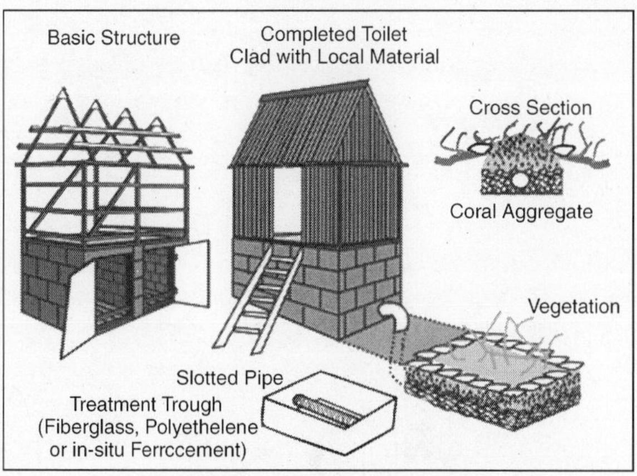

Fig. Composting Toilet

Pour Flush Toilet

A pour flush toilet has a water seal. The problems associated with odour and insects are avoided by having the water seal. Excreta deposited in the latrine pan is flushed by pouring 2 to 3 L of water into it. The mixture is directed into a pit in the same way as for a pit latrine. The processes of biodegradation of the organic wastes in the pit are exactly the same. More water percolates through the soil surrounding the pit, and the potential for groundwater pollution is higher. A pour flush toilet with a pit is therefore not suitable when groundwater table is close to the surface.

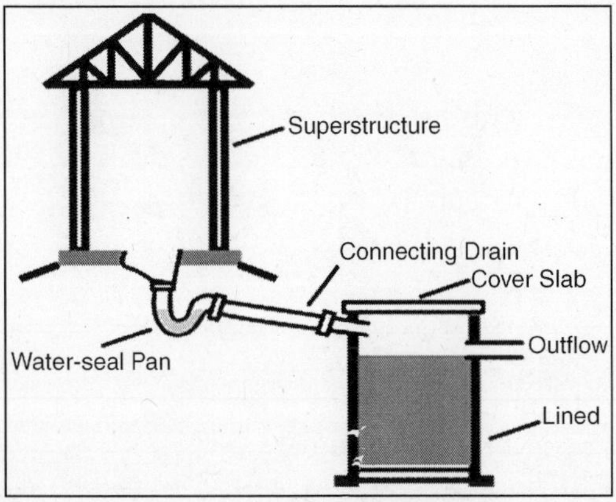

Sludge has to be regularly emptied from the pit. The use of two adjoining pits alternately enables the sludge in a full pit to undergo further decomposition while the other pit is being used, and enables manual sludge emptying after further sludge decomposition. With the use of the pit latrine, composting toilet and pour flush latrine, greywater (sullage) has to be separately treated. Greywater can be reused directly or after treatment. Disposal of greywater on-site is by use of a leach pit or trench. Limitations of disposal of greywater by leach pit or trench are similar to those applicable to septic tank.

Septic Tank

A septic tank is a watertight tank, usually located just below ground, and receives both blackwater and greywater. It can be used with pour flush toilets or cistern flush toilets. It functions as a storage tank for settled solids and floating materials (*e.g.* oils and grease). The storage time of the wastewater in the tank is usually between 2 and 4 days. About 50 per cent removal of BOD and Suspended Solids (SS) is usually achieved in a properly operated septic tank due to the settling of the solids during wastewater storage.

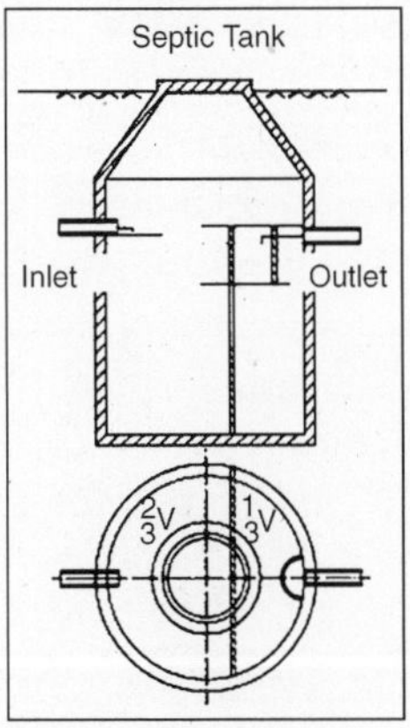

Fig. Septic Tank

A septic tank can be constructed of bricks and mortar and rendered, or of concrete. Its shape can be rectangular or cylindrical. A septic tank can be partitioned into two chambers to reduce flow short-circuiting and improve solids removal. The overflow from a septic tank is directed to a leach pit or trench. A leach pit is similar to the pit of a pit latrine or pour flush latrine. The pit must be sized to allow percolation of the volume of wastewater generated. A pit works well in soils with high permeability. In soils with lower permeability a trench can provide the larger surface area of percolation. The trench is usually filled with gravel and a distribution pipe for the wastewater is placed in this gravel layer. Soil is then placed above this gravel layer to the ground surface.

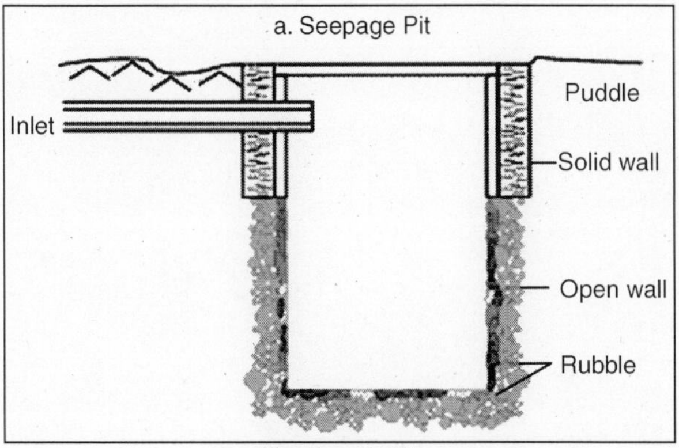

Fig. Leach Pit

A leach pit or trench does not work when the soil permeability is too low. In regions where annual evaporation is high, trees and shrubs can be used to help pump the water into the atmosphere by evapotranspiration. An evapotranspiration bed can be designed similar to a leach trench, but a suite of suitable local vegetation species tolerant of high nutrients and water are planted above and surrounding the trench. The trench should be sized to store water during the rainy season or low evaporation periods. A leach pit or drain does not work either when the groundwater table is close to ground surface. In this case off-site disposal is necessary using a settled sewerage system. If the groundwater table is not too close, an inverted leach drain as described under Improved On-site Units below can be used. The organic solids in a septic tank undergo anaerobic bacterial decomposition just as in the pit of a pit latrine. The sludge needs emptying, and the period between emptying is usually designed to be between 3 to 5 years. The sludge has to be further treated before reuse or disposal.

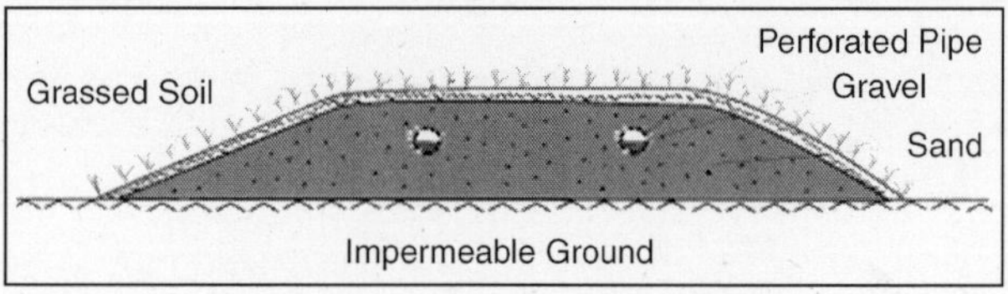

Fig. Evapotranspiration Bed

The septic tank overflow undergoes further bacterial decomposition as it percolates through a leach pit or trench. Soil bacteria, usually under aerobic conditions undertake the decomposition. The BOD of the wastewater can reach a low figure (<20 mg/L) if the distance between the bottom of the pit or trench to the groundwater table is greater than 2 m. Nutrients are not significantly removed by the bacteria and usually pollute the groundwater. Pathogenic bacteria are removed by die-off or filtration by the soil, but viruses may travel further in the soil or groundwater.

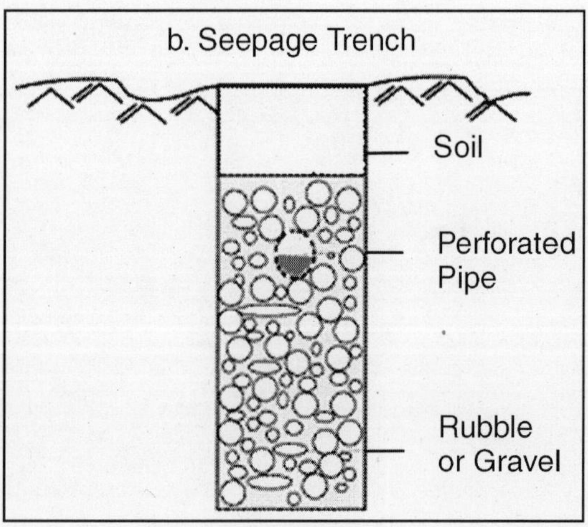

Fig. Leach Trench for Disposal of Septic Tank Effluent

Percolation of septic tank overflow is much slower compared to rainwater percolation. This is because a layer of bacterial slime grows on the surfaces of the soil particles, restricting flow. Two leach pits or trenches used alternately, say every 6 months, are better than a single leach pit or trench of the same total area for percolation, because as one is used the other will recover its percolation rate.

Improved on-site Treatment Units

Improved on-site treatment units refer to treatment units that improve the performance of one of the above on-site systems, for reducing BOD, SS and/or nutrients. Two designs are described to illustrate the main principles used. A principal aim of the improvements is to prevent groundwater pollution or enable water reuse of the treated wastewater on-site. Many designs are available using similar principles.

Inverted Trench

In the system illustrated in Figure, a plastic or impermeable liner underlies the trench of the septic tank. The liner is filled with sand or a fairly permeable soil. Overflow from the septic tank is introduced at the base of the sand layer. It flows up through the sand layer and flows over into the surrounding soil. The sand layer acts as a slow sand filter, where bacteria growing on the surfaces of the sand particles degrade the organic substances to reduce BOD.

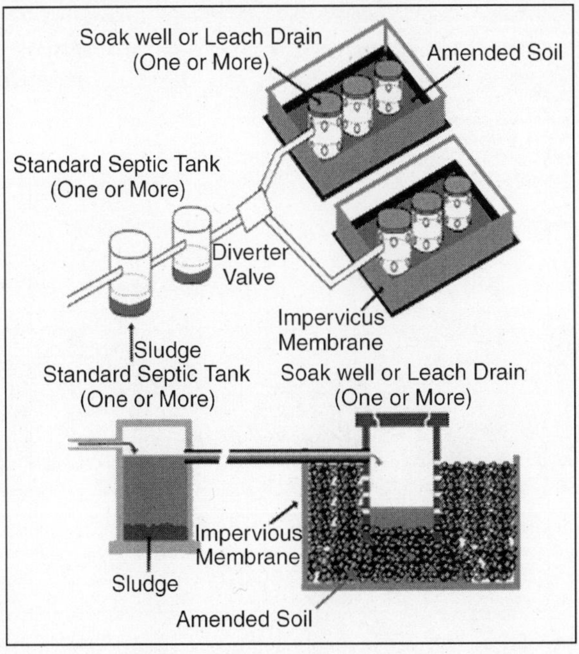

Fig. Inverted Trench (Ecomax)

Because of the fluctuating flow of wastewater with peak flows in the morning and in the evening, the upper region of the sand layer alternates between aerobic and anaerobic conditions. Under these conditions a significant part of nitrogen in the wastewater can be removed by nitrification (bacterial conversion of ammonium in the wastewater to nitrate under aerobic conditions) and denitrification (bacterial conversion of nitrate to nitrogen gas under anaerobic conditions. In addition if materials that can remove phosphate are mixed with the sand, phosphorus in the wastewater is also removed. One material, that has been found to remove phosphate effectively with a capacity for phosphorus removal for several years, is bauxite refining residue (red mud).

Aerobic Treatment Unit

An aerated treatment unit consists of a tank similar to a septic tank. The tank is partitioned into four compartments. The first compartment receives the wastewater and acts as a sedimentation tank for solids. The overflow from the first compartment goes to an aeration compartment. The aeration compartment is fitted with corrugated plastic sheets to enable bacteria to attach themselves. The aeration supplies oxygen to the bacteria decomposing the organic matter in the wastewater thus reducing its BOD. After aeration, the wastewater passes to a third compartment which acts as a second sedimentation tank. Sludge from this second sedimentation tank is pumped to the first compartment for storage. After sedimentation the wastewater overflows to a fourth compartment for storage and pumping, usually for irrigation of garden beds. If required, chlorine is applied by inserting chlorine tablets in the pipe between the third and fourth compartments. Chlorination is required when sprinklers irrigate the treated wastewater. Sub-surface irrigation is preferable, because it does not require chlorination.

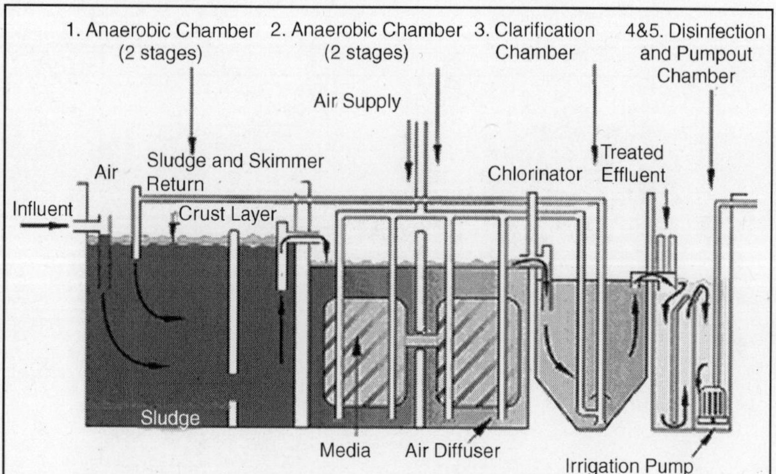

Fig. Aerated Treatment Unit (Biomax)

Power is required for aeration and pumping. For a system serving a household of up to 10 persons, the power supply rating needed is 100 W (2.5 kWh per day). This on-site unit is a miniature of an activated sludge treatment plant usually used for centralised treatment. One difference is that surfaces are provided in the aeration tank to retain bacteria during peak flows. The other difference is that sludge from the second sedimentation tank is returned to first tank for storage.

5

Environmental Biotechnology of Bioremediation

Bioremediation is a key area of 'white' biotechnology, because the elimination of a wide range of pollutants from water and soils is an absolute requirement for sustainable development. There are numerous processes of cleaning water, industrial effluents and solid wastes, using microorganisms aerobically and anaerobically. Some of them are quite sophisticated, while others are simple and adapted to the conditions of developing countries.

For instance, using microalgae (and in particular blue-green algae or cyanobacteria) in ponds to eliminate nitrogen and phosphorous, after organic matter has been degraded by bacteria, leads to water that can be recycled for irrigating non-food crops (*e.g.* cotton) or for industrial purposes; in addition, microalgal biomass can be used as feed.

Removal of Toxic Metals

Mercury is a highly toxic metal which, once released into water, accumulates in the food chain, damaging fish, shrimps and poisoning people who eat them. The infamous Minamata accident (called after a town on the Japanese island of Kyushu where the inhabitants suffered the toxic effects of fish poisoned by mercury-rich industrial effluents) is an example of the devastating effects of mercury on the central nervous system. Existing techniques of mercury removal, such as precipitation or ion exchange, are expensive and not sufficiently efficient, as small but significant amounts of mercury still remain in the water.

Researchers discovered that many bacteria had developed high tolerance to heavy metals, which related to the binding of these metals to proteins, *e.g.* metallo-thionein that binds mercury. As naturally thriving mercury-tolerant bacteria are rare and cannot be grown easily in culture, researchers at Cornell University, Ithaca, New York, inserted the metallo-thionein gene into Escherichia coli. A sufficiently large number of genetically engineered bacteria could thus treat mercury-polluted water inside a bioreactor. The efficiency of the procedure was high, as mercury was removed from polluted water down to a few nanograms per litre. Once the bacteria died, they were incinerated to recuperate the accumulated pure mercury. Mercury emissions were predicted to increase by 30 per cent throughout Europe between 1990 and 2010. The European Commission funded a demonstration project to show the feasibility and profitability of the microbial remediation technology under real time conditions.

A plant was set up at Usti-nad-Labem in the Czech Republic, and has been operating since July 2000. While phycoremediation is bioremediation based on the use of micro- and macro-algae, phytoremediation, relies on higher plants to clean water and soils from heavy metals and other pollutants, or to recolonize former mining areas (*e.g.* in South Africa, Australia, USA, Canada, France, etc.). For instance, heavy metals in industrial effluents can be concentrated in aquatic plants and thereafter recovered. In 2003 in Western Europe, there were about 1.4 million sites discovered to be polluted. Current remediation techniques are chemical or physico-chemical extraction techniques; they are costly and destroy soil structure. Phytoremediation techniques use "hyper-accumulating plants" which can store 10 to 500 more pollutants in their leaves and stems, which are thereafter harvested, incinerated, and metals recovered from ashes and reused in metallurgy. The survey of hyper-accumulating plants started by the early 1990s. They are often small plants, such as Alyssum murale, which grows on metamorphic rocks, Brassica juncea, the Indian mustard, which extracts lead, or Thlaspi, which accumulates zinc and nickel. About 400 species have been identified, including 300 that accumulate only nickel.

An endemic tree in New Caledonia, Sebertia acuminata, contains up to 20 per cent of nickel in its sap and is coloured in green (nickel is generally toxic to plants at a concentration of 0.005 per cent). In the Democratic Republic of Congo, the number of plants accumulating copper and cobalt is highest: 24 and 26 species, respectively. The accumulation efficiency is not generally very high. For some metals like silver, mercury and arsenium, there are yet no plants known to accumulate them. However, in 2000, the team of Lena Ma of the University of Florida, Gainesville, identified a fern, Pteris vittata, which tolerates and accumulates arsenium, while conserving a very rapid growth and a high biomass. Edenspace, a company from Virginia specialized in phytoremediation, acquired the rights to commercialize the fern by signing an exclusive license agreement in 2000 with the University of Florida which patented the use of the fern in phytoremediation. In the US, seven or eight similar companies in phytoremediation were already in existence in 2002, where the value of the potential market for phytoremediation was estimated at $100 million.

Bioremediation and Genetic Engineering

The molecular basis of heavy metal accumulation is being studied with a view to transferring the relevant genes to plant species having a wider geographic and ecological distribution. Transgenesis applied to phytoremediation is certainly incipient. Its application on a large scale is confronted with the evaluation of risks relating to the transfer of the bacterial transgenes to plants consumed by herbivorous animals that might acquire the property of hyper-accumulating toxic metals or compounds.

Genetic transformation of the microorganisms involved in bioremediation could enhance the process through the introduction of genes controlling specific degradation pathways; it also aims at degrading recalcitrant compounds such as pesticides and other xenosubstances. A team of US researchers led by Richard Meagher of the University of Georgia, Athens, were able to introduce into the genome of Arabidopsis thaliana, two foreign genes from Escherichia coli for the synthesis of two enzymes: one which catalyzes the transformation of arsenate into arsenite, the other which induces the formation of a complex with arsenite, that is retained in the leaves. These remarkable results led to transgenic Arabidopsis plants which could accumulate three to four times more arsenium than normal plants and could become very effective depolluting agents. The same research team is well known for their "transgenic" work particularly in the transfer to plants of bacterial genes, coding for the conversion of toxic mercury into its less dangerous volatile form. In Cambridge, United Kingdom, Neil Bruce and his team are transforming tobacco plants with bacterial genes that control the synthesis of an enzyme detoxifying TNT (trinitrotoluene). Other researchers are studying the degradation of trichloroethylene by transgenic plants.

THE ROLE OF ENVIRONMENTAL BIOTECHNOLOGY

While pharmaceutical biotechnology represents the glamorous end of the market, environmental applications are decidedly more in the Cinderella mould. The reasons for this are fairly obvious. The prospect of a cure for the many diseases and conditions currently promised by gene therapy and other biotech-oriented medical miracles can potentially touch us all. Our lives may, quite literally, be changed. Environmental biotechnology, by contrast, deals with far less apparently dramatic topics and, though their importance, albeit different, may be every bit as great, their direct relevance is far less readily appreciated by the bulk of the population. Cleaning up contamination and dealing rationally with wastes is, of course, in everybody's best interests, but for most people, this is simply addressing a problem which they would rather had not existed in the first place.

Even for industry, though the benefits may be noticeable on the balance sheet, the likes of effluent treatment or pollution control are more of an inevitable obligation than a primary goal in themselves. In general, such activities are typically funded on a distinctly limited budget and have traditionally been viewed as a necessary inconvenience. This is in no way intended to be disparaging to industry; it simply represents commercial reality. In many respects, there is a logical fit between this thinking and the aims of environmental biotechnology. For all the media circus surrounding the grand questions of our age, it is easy to forget that not all forms of biotechnology involve xenotransplantation, genetic modifica-tion, the use of stem cells or cloning. Some of the potentially most beneficial uses of biological engineering and which may touch the lives of the majority of people, however indirectly, involve much simpler approaches. Less radical and showy, certainly, but powerful tools, just the same. Environmental biotechnology is fundamentally rooted in waste, in its various guises, typically being concerned with the remediation of contamination caused by previous use, the impact reduction of current activity or the control of pollution.

Thus, the principal aims of this field are the manufacture of products in environmentally harmonious ways, which allow for the minimisation of harmful solids, liquids or gaseous outputs or the clean-up of the residual effects of earlier human occupation. The means by which this may be achieved are essentially two-fold. Environmental biotechnologists may enhance or optimise conditions for existing biological systems to make their activities happen faster or more efficiently, or they resort to some form of alteration to bring about the desired outcome. The variety of organisms which may play a part in environmental applications of biotechnology is huge, ranging from microbes through to trees and all are utilised on one of the same three fundamental bases – accept, acclimatise or alter. For the vast majority of cases, it is the former approach, accepting and making use of existing species in their natural, unmodified form, which predominates.

The Scope for Use

There are three key points for environmental biotechno-logy interventions, namely in the manufacturing process, waste management or pollution control. Accordingly, the range of businesses to which environmental biotechnology has potential relevance is almost limitless. One area where this is most apparent is with regard to waste. All commercial operations generate waste of one form or another and for many, a proportion of what is produced is biodegradable. With disposal costs rising steadily across the world, dealing with refuse constitutes an increasingly high contribution to overheads. Thus, there is a clear incentive for all businesses to identify potentially cost-cutting approaches to waste and employ them where possible.

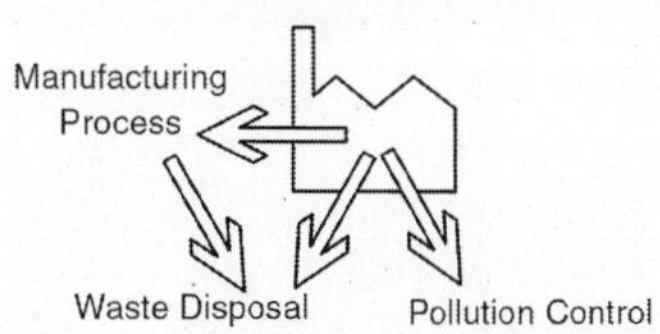

Fig. The Three Intervention Points

Changes in legislation throughout Europe, the US and elsewhere, have combined to drive these issues higher up the political agenda and biological methods of waste treatment have gained far greater acceptance as a result. For those industries with particularly high biowaste production, the various available treatment biotechnologies can offer considerable savings. Manufacturing industries can benefit from the applications of whole organisms or isolated biocomponents. Compared with conventional chemical processes, microbes and enzymes typically function at lower temperatures and pressures. The lower energy demands this makes leads to reduced costs, but also has clear benefits in terms of both the environment and workplace safety. Additionally, biotechno-logy can be of further commercial significance by converting low-cost organic feedstocks into high value products or, since enzymatic reactions are more highly specific than their chemical counterparts, by deriving final substances of high relative purity.

Almost inevitably, manufacturing companies produce wastewaters or effluents, many of which contain biodegrada-ble contaminants, in varying degrees. Though traditional permitted discharges to sewer or watercourses may be adequate for some, other industries, particularly those with recalcitrant or highly concentrated effluents, have found significant benefits to be gained from using biological treatment methods themselves on site. Though careful monitoring and process control are essential, biotechnology stands as a particularly cost-effective means of reducing the pollution potential of wastewater, leading to enhanced public relations, compliance with environmental legislation and quantifiable cost-savings to the business. Those involved in processing organic matter, for example, or with drying, printing, painting or coating processes, may give rise to the release of volatile organic compounds (VOCs) or odours, both of which represent environmental nuisances, though the former is more damaging than the latter. For many, it is not possible to avoid producing these emissions altogether, which leaves treating them to remove the offending contaminants the only practical solution. Especially for relatively low concentrations of readily water-soluble VOCs or odorous chemicals, biological technologies can offer an economic and effective alternative to conventional methods.

The use of biological cleaning agents is another area of potential benefit, especially where there is a need to remove oils and fats from process equipment, work surfaces or drains. Aside from typically reducing energy costs, this may also obviate the need for toxic or dangerous chemical agents. The pharmaceutical and brewing industries, for example, both have a long history of employing enzyme-based cleaners to remove organic residues from their process equipment. In addition, the development of effective biosensors, powerful tools which rely on biochemical reactions to detect specific substances, has brought benefits to a wide range of sectors, including the manufacturing, engineering, chemical, water, food and beverage industries.

With their ability to detect even small amounts of their particular target chemicals, quickly, easily and accurately, they have been enthusiastically adopted for a variety of process monitoring applications, particularly in respect of pollution assessment and control.

Contaminated land is a growing concern for the construction industry, as it seeks to balance the need for more houses and offices with wider social and environmental goals. The reuse of former industrial sites, many of which occupy prime locations, may typically have associated planning conditions attached which demand that the land be cleaned up as part of the development process. With urban regeneration and the reclamation of 'brown-field' sites increasingly favoured in many countries over the use of virgin land, remediation has come to play a significant role and the industry has an ongoing interest in identifying cost-effective methods of achieving it. Historically, much of this has involved simply digging up the contaminated soil and removing it to landfill elsewhere.

Bioremediation technologies provide a competitive and sustainable alternative and in many cases, the lower disturbance allows the overall scheme to make faster progress. As the previous brief examples show, the range of those which may benefit from the application of biotechnology is lengthy and includes the chemical, pharmaceutical, water, waste management and leisure industries, as well as manufacturing, the military, energy generation, agriculture and horticulture.

Clearly, then, this may have relevance to the viability of these ventures and, as was mentioned at the outset, biotechno-logy is an essentially commercial activity. Environmental biotechnology must compete in a world governed by the best practicable environmental option (BPEO) and the best available techniques not entailing excessive cost (Batneec). Consequently, the economic aspect will always have a large influence on the uptake of all initiatives in environmental biotechnology and, most particularly, in the selection of methods to be used in any given situation. It is impossible to divorce this context from the decisionmaking process. By the same token, the sector itself has its own implications for the wider economy.

THE MARKET FOR ENVIRONMENTAL BIOTECHNOLOGY

The UK's Department of Trade and Industry estimated that 15–20 per cent of the global environmental market in 2001 was biotech-based, which amounted to about 250–300 billion US dollars and the industry is projected to grow by as much as ten-fold over the following five years. This expected growth is due to greater acceptance of biotechnology for clean manufacturing applications and energy production, together with increased landfill charges and legislative changes in waste management which also alter the UK financial base favourably with respect to bioremediation. Biotechnology-based methods are seen as essential to help meet European Union (EU) targets for biowaste diversion from landfill and reductions in pollutants. Across the world the existing regulations on environmental pollution are predicted to be more rigorously enforced, with more stringent compliance standards implemented. All of this is expected to stimulate the sales of biotechnology-based environmental processing methods significantly and, in particular, the global market share is projected to grow faster than the general biotech sector trend, in part due to the anticipated large-scale EU aid for environmental clean-up in the new accession countries of Eastern Europe.

Other sources paint a broadly similar picture. The BioIndustry Association (BIA) survey, Industrial Markets for UK Biotechnology – Trends and Issues, published in 1999 does not quote any monetary sector values per year, but gives the size of the UK sector as employing 40 000 people in 1998 with an average yearly growth over 1995–98 of 20 per cent. Environmental biotech is reported as representing around 10 per cent of this sector. An Arthur Anderson report of 1997 gives the turnover of the UK biotech sector as 702 million pounds sterling in 1995/96, with a 50 per cent growth over three years. A 1998 Ernst and Young report on the European Life Sciences Sector says that the market for biotechnology products has the potential to reach 100 billion pounds sterling worldwide by 2005.

The Organisation for Economic Cooperation and Development (OECD) estimates that the global market for environmental biotechnology products and services alone will rise to some US$75 billion by the year 2000, accounting for some 15 to 25 per cent of the overall environmental technology market, which has a growth rate estimated at 5.5 per cent per annum. The UK potential market for environmental biotechnology products and services is estimated at between 1.65 and 2.75 billion US dollars and the growth of the sector stands at 25 per cent per annum, according to the Bio- Commerce Data European Biotechnology Handbook.

An unsourced quote found on a Korean University website says that the world market size of biotechnology products and services was estimated to be approximately 390 billion US dollars in the year 2000. The benefits are not, however, confined to the balance sheet. The Organisation for Economic Cooperation and Development (OECD 2001) concluded that the industrial use of biotechnology commonly leads to increasingly environmentally harmonious processes and additionally results in lowered operating and/or capital costs. For years, industry has appeared locked into a seemingly unbreakable cycle of growth achieved at the cost of environmental damage. The OECD investigation provides what is probably the first hard evidence to support the reality of biotechnology's long-heralded promise of alternative production methods, which are ecologically sound and economically efficient. A variety of industrial sectors including pharmaceuticals, chemicals, textiles, food and energy were examined, with a particular emphasis on biomass renewable resources, enzymes and bio-catalysis. While such approaches may have to be used in tandem with other processes for maximum effectiveness, it seems that their use invariably leads to reduction in operating or capital costs, or both. Moreover, the research also concludes that it is clearly in the interests of governments of the developed and developing worlds alike to promote the use of biotechnology for the substantial reductions in resource and energy consumption, emissions, pollution and waste production it offers. The potential contribution to be made by the appropriate use of biotechnology to environmental and economic sustainability would seem to be clear.

The upshot of this is that few biotech companies in the environmental sector perceive problems for their own business development models, principally as a result of the wide range of businesses for which their services are applicable, the relatively low market penetration to date and the large potential for growth. Competition within the sector is not seen as a major issue either, since the field is still largely open and unsaturated. Moreover, there has been a discernible tendency in recent years towards niche specificity, with companies operating in more specialised subarenas within the environmental biotechnology umbrella. Given the number and diversity of such possible slots, coupled with the fact that new opportunities and the technologies to capitalise on them, are developing apace, this trend seems likely to continue.

It is not without some irony that companies basing their commercial activities on biological organisms should themselves come to behave in such a Darwinian fashion. However, the picture is not entirely rosy. Typically the sector comprises a number of relatively small, specialist companies and the market is, as a consequence, inevitably fragmented. Often the complexities of individual projects make the application of 'standard' off-theshelf approaches very difficult, the upshot being that much of what is done must be significantly customised. While this, of course, is a strength and of great potential environmental benefit, it also has hard commercial implications which must be taken into account.

A sizeable proportion of companies active in this sphere, have no products or services which might reasonably be termed suitable for generalised use, though they may have enough expertise, experience or sufficiently perfected techniques to deal with a large number of possible scenarios. The fact remains that one of the major barriers to the wider uptake of biological approaches is the high perceived cost of these applications.

Part of the reason for this lies in historical experience. For many years, the solutions to all environmental problems were seen as expensive and for many, particularly those unfamiliar with the multiplicity of varied technologies available, this has remained the prevalent view. Generally, there is often a lack of financial resource allocation available for this kind of work and biotech providers have sometimes come under pressure to reduce the prices for their services as a result.

Greater awareness of the benefits of biotechnology, both as a means to boost existing markets and for the opening up of new ones, is an important area to be addressed. Many providers, particularly in the UK, have cited a lack of marketing expertise as one of the principal barriers to their exploitation of novel opportunities.

In addition, a lack of technical understanding of biotech approaches amongst target industries and, in some cases, downright scepticism regarding their efficacy, can also prove problematic. Good education, in the widest sense, of customers and potential users of biological solutions will be one major factor in any future upswing in the acceptance and utilisation of these technologies.

Modalities and Local Influences

Another of the key factors affecting the practical uptake of environmental biotechnology is the effect of local circumstances. Contextual sensitivity is almost certainly the single most important factor in technology selection and represents a major influence on the likely penetration of biotech processes into the marketplace.

Neither the nature of the biological system, nor of the application method itself, play anything like so relevant a role. This may seem somewhat unexpected at first sight, but the reasons for it are obvious on further inspection. While the character of both the specific organisms and the engineering remain essentially the same irrespective of location, external modalities of economics, legislation and custom vary on exactly this basis. Accordingly, what may make abundant sense as a biotech intervention in one region or country, may be totally unsuited to use in another. In as much as it is impossible to discount the wider global economic aspects, disassociating political, fiscal and social conditions equally cannot be done, as the following example illustrates.

In 1994, the expense of bioremediating contaminated soil in the United Kingdom greatly exceeded the cost of removing it to landfill. Six years later, with successive changes of legislation and the imposition of a landfill tax, the situation has almost completely reversed. In those other countries where landfill has always been an expensive option, remediation has been embraced far more readily. While environmental biotechnology must, inevitably, be viewed as contextually dependent, as the previous example shows, contexts can change. In the final analysis, it is often fiscal instruments, rather than the technologies, which provide the driving force and sometimes seemingly minor modifica-tions in apparently unrelated sectors can have major ramifications for the application of biotechnology.

The legal framework is another aspect of undeniable importance in this respect. Increasingly tough environmental law makes a significant contribution to the sector and changes in regulatory legislation are often enormously influential in boosting existing markets or creating new ones.

When legislation and economic pressure combine, as, for example, they have begun to do in the European Landfill Directive, the impetus towards a fundamental paradigm shift becomes overwhelming and the implications for relevant biological applications can be immense. There is a natural tendency to delineate, seeking to characterise technologies into particular categories or divisions. However, the essence of environmental biotechnology is such that there are many more similarities than differences. Though it is, of course, often helpful to view individual technology uses as distinct, particularly when considering treatment options for a given environmental problem, there are inevitably recurrent themes which feature throughout the whole topic. Moreover, this is a truly applied science. While the importance of the laboratory bench cannot be denied, the controlled world of research translates imperfectly into the harsh realities of commercial implementation. Thus, there can often be a dichotomy between theory and application and it is precisely this fertile ground which is explored in the present work.

In addition, the principal underlying approach of specifically environmental biotechnology, as distinct from other kinds, is the reliance on existing natural cycles, often directly and in an entirely unmodified form. Thus, this science stands on a foundation of fundamental biology and biochemistry. To understand the application, the biotechnologist must simply examine the essential elements of life, living systems and ecological circulation sequences. However engineered the approach, this fact remains true. In many respects, environmental biotechnology stands as the purest example of the newly emergent bioindustry, since it is the least refined, at least in terms of the basis of its action.

In essence, all of its applications simply encourage the natural propensity of the organisms involved, while seeking to enhance or accelerate their action. Hence, optimisation, rather than modification, is the typical route by which the particular desired end result, whatever it may be, is achieved and, consequently, a number of issues feature as common threads of individual technologies.

Integrated Approach

ntegration is an important aspect for environmental biotechnology. One theme that will be developed throughout this book is the potential for different biological approaches to be combined within treatment trains, thereby producing an overall effect which would be impossible for any single technology alone to achieve. However, the wider goal of integration is not, of necessity, confined solely to the specific methods used. It applies equally to the underpinning knowledge that enables them to function in the first place and an understanding of this is central to the rationale behind this book. In some spheres, traditional biology has become rather unfashionable and the emphasis has shifted to more exciting sounding aspects of life science.

While the new-found concentration on 'Ecological processes', or whatever, sounds distinctly more 'Environ-mental', in many ways and somewhat paradoxically, it sometimes serves the needs of environmental biotechnology rather less well. The fundamentals of living systems are the stuff of this branch of science and, complex though the whole picture may be, at its simplest the environmental biotechnologist is principally concerned with a relatively small number of basic cycles. In this respect, a good working knowledge of biological processes like respiration, fermentation and photosynthesis, a grasp of the major cycles by which carbon, nitrogen and water are recycled and an appreciation of the flow of energy through the biosphere must be viewed as prerequisites.

The intent here has been neither to insult the readership by parading what is already well known, nor gloss over aspects which, if left unexplained, at least in reasonable detail, might only serve to confuse. However, this is expressly not designed to be a substitute for much more specific texts on these subjects, nor an entire alternative to a cohesive course on biology or biochemistry. The intention is to introduce and explain the necessary aspects and elements of various metabolic pathways, reactions and abilities as required to advance the reader's understanding of this particular branch of biotechnology.

A large part of the reasons for approaching the subject in this way is the fact that there really is no such thing as a 'typical' environmental biotechnologist. Practitioners come into the profession from a wide variety of disciplines and by many different routes. Thus, amongst their ranks are agronomists, biochemists, biologists, botanists, enzymologists, geneticists, microbiologists, molecular biologists, process engineers and protein technologists, all of whom bring their own particular skills, knowledge base and experiences. The applied nature of environmental biotechnology is obvious. While the science underlying the processes themselves may be as pure as any other, what distinguishes this branch of biological technology are the distinctly real-life purposes to which it is put. Hence, part of the intended function of this book is to attempt to elucidate the former in order to establish the basis of the latter.

At the same time, as any applied scientist will confirm, what happens in the field under operational conditions represents a distinct compromise between the theoretical and the practically achievable. At times, anything more than an approximation to the expected results may be counted as something of a triumph of environmental engineering.

Closing Remarks

The celebrated astronomer and biologist, Sir Fred Hoyle, said that the solutions to major unresolved problems should be sought by the exploration of radical hypotheses, while simultaneously adhering to well-tried and tested scientific tools and methods. This approach is particularly valid for environmental biotechnology. With new developments in treatment technologies appearing all the time, the list of what can be processed or remediated by biological means is ever changing. By the same token, the applications for which biotechnological solutions are sought are also subject to alteration. For the biotech sector to keep abreast of these new demands it may be necessary to examine some truly 'radical hypotheses' and possibly make use of organisms or their derivatives in ways previously unimagined. This is the basis of innovation; the inventiveness of an industry is often a good measure of its adaptability and commercial robustness.

BIOREMEDIATION USE OF BIOLOGICAL SYSTEMS

Bioremediation is the use of biological systems for the reduction of pollution from air or from aquatic or terrestrial systems. Microorganisms and plants are the biological systems which are generally used. Biodegradation with micro-organisms is the most frequently occurring bioremediation option. Micro-organisms can break down most compounds for their growth and/or energy needs. These biodegradation processes may or may not need air. In some cases, metabolic pathways which organisms normally use for growth and energy supply may also be used to break down pollutant molecules.

In these cases, known as co-metabolism, the microorganism does not benefit directly. Researchers have taken advantage of this phenomenon and use it for bioremediation purposes. A complete biodegradation results in detoxification by mineralising pollutants to carbon dioxide, water and harmless inorganic salts. Incomplete biodegradation will yield breakdown products which may or may not be less toxic than the original pollutant. Incomplete biodegradation of tri- or tetrachloroethylene for instance can yield vinylchloride, which is more toxic and carcinogenic than the original compounds.

Biodegradation may occur spontaneously, in which case the expressions "intrinsic bioremediation" or "natural attenuation" are often used. In many cases however the natural circumstances are not favourable enough for this to happen due to the lack of enough nutrients, oxygen or suitable bacteria. Such situations may be improved by supplementing one or more of these prerequisites. Extra nutrients were for instance disseminated to speed up the breakdown of the oil spilled on 1000 miles of Alaskan shoreline by the super tanker Exxon Valdez in 1989. The future trend in bioremediation increasingly is to look at the speed of unaided biodegradation first and act only if there is insufficient activity to remove the pollutant quickly enough to prevent any expected risks of the pollutant. Bioremediation techniques can be used to reduce or to remove hazardous waste which has already polluted the environment. They can also be used to treat waste streams before they leave production facilities: end-of-pipeprocesses.

WASTE WATER AND INDUSTRIAL EFFLUENTS

Micro-organisms in sewage treatment plants remove the more common pollutants from waste water before it is discharged into rivers or the sea. Increasing industrial and agricultural pollution has led to a greater need for processes that remove specific pollutants such as nitrogen and phosphorus compounds, heavy metals and chlorinated compounds. New methods include aerobic, anaerobic and physico-chemical processes in fixed-bed filters and in bioreactors in which the materials and microbes are held in suspension.

The costs of waste water treatment can be reduced by the conversion of wastes into useful products. For example, heavy metals and sulphur compounds can be removed from waste streams of the galvanisation industry by the aid of sulphur metabolising bacteria and reused. Another example is the production of animal feed from the fungal biomass which remains after the production of penicillin. Most anaerobic waste water treatment systems produce useful biogas.

Drinking and Process Water

Abundant supplies of water are vital for modern urban and industrial development. By the turn of this century, it is estimated that two-thirds of the world's nations will be water stressed - using clean water faster than it is replenished in aquifers or rivers. A very important aspect of biotechnology is therefore its potential for the reclamation and purification of waste waters for re-use. Public concern has also increased over the quality of drinking water.

Not only does water need to be recycled in the development of sustainable use of resources, overall quality must also be improved to satisfy consumers. In many agricultural regions of the world, animal wastes and excess fertilisers result in high levels of nitrates in drinking water. Biotechnology has provided successful methods by which these compounds can be removed from processed water before it is delivered to customers.

Air and Waste Gases

Originally, industrial waste gas treatment systems were based on cheap compost-filled filters that removed odours. Such systems still exist. However, slow processing rates and the short life of such filters drove research into better methods such as bioscrubbers, in which the pollutants are washed out using a cell suspension and biotrickling filters, in which the pollutant is degraded by micro-organisms immobilised on an inert matrix and provided with an aqueous nutrient film trickling through the device.

The selection of micro-organisms that are more efficient at metabolising pollutants has also led to better air and gas purifying biofilters. Examples are a bioscrubber based system for the simultaneous removal of nitrogen and sulphur oxides from the flue gas of blastfurnaces which has been developed as an alternative to the classical limestone gypsum process, and the elimination of styrene from the waste gas of polystyrene processing industries by a biofilter containing fungi.

Soil and Land Treatment

Both in situ (in its original place) and ex situ (somewhere else) methods are commercially exploited for the cleanup of soil and the associated groundwater. In situ treatments may include the introduction of micro-organisms (bioaugmentation), ventilation and/or adding nutrient solutions (biostimulation). Ex situ treatment involves removing the soil and groundwater and treating it above ground. The soil may be treated as compost, in soil banks, or in specialised slurry bioreactors. Groundwater is treated in bioreactors and either pumped back into the ground or drained into surface water.

Bioremediation of land (biorestoration) is often cheaper than physical methods and its products are harmless if complete mineralisation takes place. Its action can however, be time-consuming, tying up capital and land. The in situ bioremediation of the ground under petrol stations has already become common practice but also for chlorinated solvents like tri- and tetrachloroethylene in situ bioremediation is possible. The applicability of in situ bioremediation is and probably will remain dependent on the physical parameters of the soil, mainly its transport properties. Bioremediation using plants is called phytoremediation.

This technique is presently already used to remove metals from contaminated soils and groundwater and is being further explored for the remediation of other pollutants. The combined use of plants and bacteria may also be possible. Certain bacteria live closely associated with the roots of plants and depend on substances excreted by the roots. Such rhizobacteria, whose numbers are much higher than those of other soil bacteria, may be genetically modified to break down pollutants. Research is being conducted to test this hypothesis.

Solid Waste

Domestic solid wastes are a major problem in our consumption society. Their elimination is both costly and warrants constant surveillance in terms of groundwater and air pollution. Yet, for a major part they are composed of readily biodegradable organics. In this respect, source separated bio-wastes can be converted to a valuable resource by composting or anaerobic digestion. In recent years, both processes have seen remarkable developments in terms of process design and control.

Particularly, anaerobic digestion of solid wastes in high-rate anaerobic digesters has gained increasing public acceptance because it permits the recovery of substantial amounts of high-value biogas together with a high quality stable organic residue and this without giving rise to environmental nuisance. Moreover, anaerobic digestion of mixed solid wastes is under intensive development because in the near future it may be an important step in recycling of solid wastes and constitute an alternative to incineration.

Prevention

Progressively more industrial companies are developing processes with reduced environmental impact responding to the international call for the development of a sustainable society. There is a pervading trend towards less harmful products and processes; away from "end-of-pipe" treatment of waste streams. Biotechnology is pre-eminently suitable to contribute to this trend and it has already done so in many cases, both by the improvement of existing processes and by the development of new ones.

Process Improvement

Many industrial processes have been made more environmentally friendly by the use of enzymes. Enzymes are biological catalysts that are highly efficient and have numerous advantages over non-biological catalysts. They are non-toxic and biodegradable, work best at moderate temperatures and in mild conditions, and have fewer side reactions than traditional methods because they are highly specific. Production methods that employ enzymes are generally not only cleaner and safer compared with other methods, but mostly also more economic in energy and resource consumption.

Their specificity does however mean that it is not always easy to find the appropriate enzyme for a given application. Enzymes are already widely employed in industry and have been for many years. New techniques and approaches to protein design and molecular modelling are enabling researchers to develop novel enzymes active at high temperatures, in non-aqueous solvents and as solids.

Product Innovation

Biotechnology also can help to produce new products which have less impact on the environment than their predecessors. The production of new biomaterials like bioplastics avoids the use of non-renewable resources like fossil fuels. Potatoes normally contain 80 per cent amylopectin but also 20 per cent amylose which is unwanted in many applications. For the isolation of pure amylopectin large amounts of water and energy are consumed. A Dutch company has developed a genetically modified potato variety which no longer contains amylose and hence can be processed with less impact on the environment.

The use of genetically modified plant varieties that are resistant against insects and/or diseases may considerably diminish the use of pesticides which not only prevents the use of the -mostly non-renewable- raw materials, energy and labour necessary for their production but also will reduce the negative impacts of their residues. Many more of these biotechnological solutions for pollution have been developed. Future developments may involve things which at this moment seem science-fiction to most people such as the replacement of chemically produced super-fibres by microbially produced spider web silk. One thing that should not be forgotten however is that the increased use of biological systems in industry should be accompanied by adequate training and protection of workers handling these systems, just like in other sections of industry.

DETECTION AND MONITORING

Detection and Monitoring of Pollutants

A wide range of biological methods are already in use to detect pollution incidents and for the continuous monitoring of pollutants. Long established measures include: counting the number of plant, animal and microbial species, counting the numbers of individuals in those species or analysing the levels of oxygen, methane or other compounds in water. More recently, biological detection methods using biosensors and immunoassays have been developed and are now being commercialised.

Most biosensors are a combination of biological and electronic devices - often built onto a microchip. The biological component might be simply an enzyme or antibody, or even a colony of bacteria, a membrane, neural receptor, or an entire organism. Immobilised on a substrate, their properties change in response to some environmental effect in a way that is electronically or optically detectable.

It is then possible to make quantitative measurements of pollutants with extreme precision or to very high sensitivities. The sensors can be designed to be very selective, or sensitive to a broad range of compounds. For example, a wide range of herbicides can be detected in river water using algal-based biosensors; the stresses inflicted on the organisms being measured as changes in the optical properties of the plant's chlorophyll. Microbial biosensors are micro-organisms which produce a reaction upon contact with the substance to be sensed. Usually they produce light but cease to do so upon contact with substances which are toxic to them. Both naturally occurring light emitting microorganisms as well as specially developed ones are used.

Positively acting bacterial biosensors have been constructed which start emitting light upon contact (and subsequent reaction) with a specific pollutant. In the USA such a light emitting bacterium has been approved for the detection of polyhalogenated aromatic hydrocarbons in field tests. Immunoassays use labelled antibodies (complex proteins produced in biological response to specific agents) and enzymes to measure pollutant levels. If a pollutant is present, the antibody attaches itself to it; the label making it detectable either through colour change, fluorescence or radioactivity. Immunoassays of various types have been developed for the continuous, automated and inexpensive monitoring of pesticides such as dieldrin and parathion.

The nature of these techniques, the results of which can be as simple as a colour change, make them particularly suitable for highly sensitive field testing where the time and large equipment needed for more traditional testing is impractical. Their use is however limited to pollutants which can trigger biological antibodies. If the pollutants are too reactive, they will either destroy the antibody or suppress its activity and so also the effectiveness of the test.

Detection and Monitoring of Microorganisms used for Bioremediation

When laboratory grown micro-organisms are inoculated into a bioremediation site (bioaugmentation) it often becomes necessary to monitor their presence and/or multiplication to check the progress of the process. This is especially true and even required when genetically modified micro-organisms are involved. The traditional technique to detect the presence of micro-organisms in soil is direct plating on selective media.

This is greatly facilitated if the organism contains a marker which can be selected for. Newer techniques include the above mentioned immunological and light-based bioreporter techniques. The spatial distribution of specific microorganisms in a sample can be determined microscopically and non-invasively by using fluorescent in situ hybridisation (FISH) of micro-organisms. The most sensitive and specific technique is the direct isolation and amplification of DNA from soil, which is increasingly being used.

Detection and Monitoring of Ecological Effects

Bioremediation is aimed at improving the quality of the environment by removing pollutants. However, the disappearance of the original pollutant is not the only criterion by which the success of a bioremediation operation is determined. (Even more) toxic metabolites may be produced from the pollutant or the biodegrading bacterium may cause diseases or produce substances that are harmful to useful micro-organisms, plants, animals or humans.

All these negative effects, are of course, excluded as much as possible in advance by getting as familiar as possible with the organism through extensive literature searches and microcosm studies in which the bioremediation process is simulated in the laboratory. To avoid unexpected effects, especially after the release of new member of the eco-system like a genetically modified organism, the monitoring of the ecological effects of a bioremediation operation may be required. The problem with monitoring ecological effects is what to monitor. Numerous ecological effects are possible but not all of them may be relevant or permanent or even the result of the bioremediation operation. The parameters to be monitored are usually determined case-by-case. Monitoring techniques may include all of those mentioned in the two previous subsections on detection and monitoring.

PRINCIPLES OF BIOREMEDIATION

Environmental biotechnology is not a new field; composting and wastewater treatments are familiar examples of old environmental biotechnologies. However, recent studies in molecular biology and ecology offer opportunities for more efficient biological processes. Notable accomplishments of these studies include the clean-up of polluted water and land areas.

Bioremediation is defined as the process whereby organic wastes are biologically degraded under controlled conditions to an innocuous state, or to levels below concentration limits established by regulatory authorities. By definition, bioremediation is the use of living organisms, primarily microorganisms, to degrade the environmental contaminants into less toxic forms. It uses naturally occurring bacteria and fungi or plants to degrade or detoxify substances hazardous to human health and/or the environment. The microorganisms may be indigenous to a contaminated area or they may be isolated from elsewhere and brought to the contaminated site. Contaminant compounds are transformed by living organisms through reactions that take place as a part of their metabolic processes. Biodegradation of a compound is often a result of the actions of multiple organisms.

When microorganisms are imported to a contaminated site to enhance degradation we have a process known as bioaugmentation. For bioremediation to be effective, microorganisms must enzymatically attack the pollutants and convert them to harmless products. As bioremediation can be effective only where environmental conditions permit microbial growth and activity, its application often involves the manipulation of environmental parameters to allow microbial growth and degradation to proceed at a faster rate.

Like other technologies, bioremediation has its limitations. Some contaminants, such as chlorinated organic or high aromatic hydrocarbons, are resistant to microbial attack. They are degraded either slowly or not at all, hence it is not easy to predict the rates of clean-up for a bioremediation exercise; there are no rules to predict if a contaminant can be degraded. Bioremediation techniques are typically more economical than traditional methods such as incineration, and some pollutants can be treated on site, thus reducing exposure risks for clean-up personnel, or potentially wider exposure as a result of transportation accidents. Since bioremediation is based on natural attenuation the public considers it more acceptable than other technologies. Most bioremediation systems are run under aerobic conditions, but running a system under anaerobic conditions may permit microbial organisms to degrade otherwise recalcitrant molecules.

Factors of Bioremediation

The control and optimization of bioremediation processes is a complex system of many factors. These factors include: the existence of a microbial population capable of degrading the pollutants; the availability of contaminants to the microbial population; the environment factors (type of soil, temperature, pH, the presence of oxygen or other electron acceptors, and nutrients).

PHYTOREMEDIATION AND BIOTECHNOLOGY

The first goal in phytoremediation is to find a plant species which is resistant to or tolerates a particular contaminant with a view to maximising it's potential for phytoremediation. Resistant plants are usually located growing on soils with underlying metal ores or on the boundary of polluted sites. Once a tolerant plant species has been selected traditional breeding methods are used to optimize the tolerance of a species to a particular contaminant. Agricultural methods such as the application of fertilisers, chelators, and pH adjusters can be utilised to further improve the potential for phytoremediation.

Genetic modification offers a new hope for phytoremediation as GM approaches can be used to overexpress the enzymes involved in the existing plant metabolic pathways or to introduce new pathways into plants. Richard Meagher and colleagues introduced a new pathway into *Arabidopsis* to detoxify methylmercury, a common form of environmental pollutant to elemental mercury which can be volatilised by the plant.

- The genes originated in gram-negative bacteria.
- *MerB* encodes a protein organomercurial lyase converts methylmercury to ionic mercury.
- *MerA* encodes mercuric reductase, which reduces ionic mercury to the elemental form.
- Arabidopsis plants were transformed with either *MerA* or *MerB* coupled with a consitutive 35S promoter.
- The MerA plants were more tolerant to ionic mercury, volatilised elemental mercury, and were unaffected in their tolerance of methylmercury.
- The MerB Plants were significantly more tolerant to methylmercury and other organomercurials and could also convert mthylmercury to ionic mercury which is approximately 100 times less toxic to plants.
- MerA MerB double transgenics were produced in an F2 generation. These plants not only showed a greater resistance to organic mercury when compared to the MerA, MerB, and wildtype plants but also capable of volatilising mercury when supplied with methylmercury.
- The same *MerA/MerB* inserts have been used in other plant species

including tobacco(*Nicotiana tabacum*), yellow poplar(*Liriodendron tulipifera*).

- Wetland species (bulrush and cat-tail) and water tolerant trees (willow and poplar) have also been targetted for transformation.

Risk Assessment

The use of phytoremediation in the field is subject to many environmental concerns, especially in the light of the recent pulic hysteria about the release of GM crops into the environment. Even if non GM strains of plants are used there are still many concerns:

- It is unknown what ecological effects hyperaccumulator plants may have if ingested by animals.
- Fallout from senescing tissues in autumn may also re-enter the food chain.
- Do volatilized contaminants remain at 'safe' levels in the atmosphere.
- Exposure of the ecosystem to contaminants is prolonged as phytoremediation is a relatively slow process.

However there are other issues that affect the risk assesment for the use of transgenic organisms as phytoremediators. Not only do such organisms have the same risks as witld type remediators but they also have the same risks as releasing any GM organism into the field have:

- The potential genetic pollution of native species.
- Potential for the gene to recombine with other genes possibly leading to the hyperaccumulation of non-contaminant compounds.
- Reporter/marker genes may also escape into the environment
- The GM plants may revert to a wild type genotype.

TYPES OF PHYTOREMEDIATION

There are five basic types of phytoremediation, categorized by the type of material they remediate – metals or organic chemicals.

For metal Contamination

- *Phytoextraction*: In this process, the plant absorbs metal contaminants through the roots and moves them into the upper portions of the plant (stems and the leaves). Hyperaccumulators are often used in this process for their enhanced capacity to extract contaminants from their surrounding environment. The plants are then harvested and discarded appropriately. It may also be possible to recover the original metals from the plants.
- *Rhizofiltration:* This method applies specifically to surface and ground water remediation. In this process, contaminants are extracted either directly by absorption through the roots, or indirectly through root adsorption, meaning that contaminants are attracted to and held by

the roots. Plants that have been grown in clean water are transplanted to the contaminated water site. When the roots become saturated with the contaminants, they are harvested and new ones are planted. The use of constructed wetlands to treat wastewater and landfill leachate are examples of rhizofiltration.

- *Phytostabilization*: Plants are used to contain soil and water contaminants or reduce their mobility through the surrounding environment. This is accomplished by contaminant absorption or adsorption through the plant roots, or by decreasing soil erosion and wind-blown dust. For metals, which do not biodegrade, this is an effective method for preventing their spread to ground or surface water.

For Organic Chemicals

- *Phytotransformation*: In this process, the plant absorbs and breaks down organic chemicals in contaminated soil and groundwater through its metabolic processes.
- *Rhizosphere Bioremediation (or Plant Assisted Bioremediation/ Rhizodegradation)*: Plants stimulate the growth of microorganisms in the area around their roots (called the rhizosphere) through the release of natural substances. These microorganisms, such as yeast, fungi, and bacteria, degrade contaminants through their metabolic processes.

WORKING PRINCIPLE OF PHYTOREMEDIATION

Phytoremediation takes advantage of a plant's natural ability to absorb, accumulate, or metabolize contaminants from the soil or other media in which it grows. Interactions between these plants and microorganisms that live in the soil can also contribute to phytoremediation. As part of normal growth and development, plants naturally absorb and metabolize certain organic chemicals and metals. Some metals are essential for plant growth, including zinc, copper, and iron. Some plants, classified as hyperaccumulators, absorb more metal than other plants, including metals that do not appear necessary for plant function.

Harvest and Disposal of Plants

When the plants have absorbed and accumulated contaminants, they can be harvested and discarded. If organic chemical contaminants are degraded into molecules like water and carbon dioxide, the plants may not require any special method of disposal. Controlled incineration is the most common method used to dispose plants that have absorbed large amounts of contamination. This process produces ashes, which can then be discarded at appropriate waste sites. For plants that have absorbed metals, controlled incineration produces ashes with a high metal content. Researchers are working on methods to recover the original metals from these ashes.

Since phytoremediation is a technology still in the early stages of development, many disposal and metal recovery methods are still being explored. These methods include sun-drying, composting, and leaching.

Biotechnology and Phytoremediation

Many different genes are suspected to be involved in regulating metal absorption in plants. These genes control the solubility of metals in the soil surrounding the roots, as well as the transport proteins that move metals into root cells and up into the plant's shoots. Improved knowledge of the genes, molecular mechanisms, and roles microorganisms play in governing phytoremediation will allow scientists to better apply plants to the clean-up of contaminated sites.

Current Research Areas in Phytoremediation

Researchers are examining the exact mechanisms surrounding metal transport in plants, and why some plants can absorb and tolerate high amounts of toxic metals while others cannot. Identified genes are being cloned, and certain plants are being genetically modified to tolerate metal contamination. In some cases, plants are being genetically modified with bacterial genes. For example, researchers at the University of Georgia genetically modified yellow poplar trees with a gene from a mercury-resistant bacteria. In the lab, the trees thrived in mercury-heavy soils. The researchers are further studying the trees in a greenhouse setting. Scientists are also studying biodiversity at metal-contaminated sites across Canada. This work may determine whether seed banks can be established from these wild species, and whether they can be successfully grown in greenhouses. This could eventually lead to the creation of an inventory of plants that could be used at other metal-contaminated sites.

Phytoremediation and Natural Resources

The use of phytoremediation technology is being considered in various industries, including mining and oil. In 1997, workers at Saskatchewan Federal Co-operatives Limited planted 203 poplars on a test site near Kelvington in an abandoned gas station that contained underground petroleum pollution. Balsam poplars were chosen for the experiment because they were hardy, fast-growing and readily available in Saskatchewan. An adult poplar can take in up to 120 litres of water per day. Their experiment proved that the trees could survive in areas with high levels of petroleum contamination. They will measure the amount of petroleum left in the soil and expect to see results in 10 years.

Sustainable Development and Phytoremediation

In order for a technology to be sustainable, it should be economically viable and environmentally compatible. Phytoremediation uses the existing capabilities of plants and the systems they support to clean up soil and water. It is more cost-effective than traditional remediation methods for contaminated soil, which involve digging up the entire contaminated area and taking it away to another location for chemical treatment, incineration, or burial. Phytoremediation takes less labour and does not disturb the natural surroundings of the contamination site. Although phytoremediation takes time, it is a good way to make use of naturally existing resources.

Technical and Regulatory Guidance on Phytoremediation

Phytoremediation is the use of vegetation for *in situ* or *ex situ* treatment of contaminated soils, sediments, and water where green plants degrade, assimilate, metabolize, or detoxify inorganic and organic chemicals. This emerging technology has become attractive due to its relatively low cost coupled with its effectiveness and aesthetic nature of using plants to clean up sites. However, phytoremediation has not been widely utilized due to the limited knowledge on this technology.

Therefore, the Interstate Technology and Regulatory Cooperation (ITRC) formed a Phytoremediation Work Group in 1998, to identify, provide guidance, and reduce technical and regulatory barriers for applying this technology. Established in 1995, the ITRC is a state-led, national coalition of personnel from the regulatory and technology programs of more than 25 states, three federal agencies, and tribal, public, and stakeholders.

This presentation focuses on two guidance documents produced by the ITRC Work Group to explain phytoremediation to consultants, stakeholders, scientists, regulators, and the public. The first document, which is a decision tree for soil, groundwater, and sediments, provides a tool that can be used to determine the potential effectiveness of phytoremediation at a given site. The decision tree document allows the user to take the basic information from a site and through a flow chart layout, decide if phytoremediation is feasible at the specific site. Background information and a glossary are provided to support each flow chart, one corresponding to each of the media. The ITRC Phytoremediation Work

Group is also working on a second, more comprehensive document encompassing more of the technical and regulatory information and guidance for applying phytoremediation. This includes providing suggestions and identifying lessons learned from case studies in order to reduce future technical and regulatory barriers.

Phytoremediation for Hydraulic Control of Shallow Groundwater

Phytoremediation has emerged as an innovative and aesthetic remedial strategy for various environmental applications. Phytoremediation has been used for surface water infiltration control, treatment of contaminated soil, and as outlined in this case study, groundwater hydraulic control. In September

1996, a remedial action plan (RAP) was developed for implementation of a Phytobarrier in conjunction with monitored natural attenuation at an active petroleum retail station in Ohio. RAP pilot testing involved a slug test to estimate the aquifers hydraulic conductivity and a bioattenuation evaluation. The slug test indicated a low hydraulic conductivity consistent with silty clays observed at the site. The bioattenuation evaluation concluded that a substantial microbial population exists beneath the site that is actively degrading petroleum hydrocarbons under highly reducing conditions. The evaluation identified natural attenuation as a passive mechanism for source control but due to the proximity of local residents, the potential for migration of impacted groundwater needed to be addressed.

In 1997, a Phytobarrier system was installed by planting hybrid poplar trees within a trench along the western site border, downgradient of the source area. The hydrid poplar species was selected due to rooting characteristics, fast growth, and site remedial objectives. A piezometer network was installed upgradient, downgradient and within the Phytobarrier to monitor the groundwater table. During planting, soil within the Phytobarrier was aggressively amended and fertilized resulting in 8 to 12 feet of tree growth during the first growing season.

During the second growing season, tree roots were observed to a depth of 8 feet. Based on groundwater elevation data in the third growing season, hydraulic gradient across the Phytobarrier had decreased more than 3 feet compared to baseline data. This project proves that phytoremediation is a viable technology to provide groundwater hydraulic control in an aquifer with low hydraulic conductivity.

Phytoremediation of Arsenic Contaminated Soils Using A Hyperaccumulating Fern

Arsenic is of great environmental concern due to its extensive contamination and carcinogenic toxicity. There is a great need for reliable and cost-effective technologies capable of reducing arsenic in soils to environmentally acceptable levels. Phytoremediation, a plant-based green technology, has been successfully used to remove heavy metals from contaminated soils.

However, no arsenic hyperaccumulating plant is currently available until recently. We have discovered an extremely efficient arsenic-hyperaccumulating fern. A number of fern samples were collected from an arsenic contaminated soil and analyzed for arsenic concentrations. In addition, ferns after growing in artificially contaminated soils for up to 8 weeks in a greenhouse were harvested and analyzed for arsenic concentrations. The highest arsenic concentrations in the aboveground biomass in ferns growing in the arsenic contaminated soil in the field was 7,500 ppm, with arsenic concentration in aboveground biomass being up to 200 times greater than those of soils. After 4 weeks, the arsenic concentration reached over 2 per cent in the aboveground biomass of the fern growing in the soil spiked with 500 ppm arsenic. Obviously, this fern has an extraordinary capability to uptake a large quantity of arsenic from soils and translocate them to aboveground biomass. This fern has a great potential to be used for phytoremediating arsenic contaminated soils and wastes.

AEROBIC BIODEGRADATION OF ORGANIC POLLUTANTS

The burgeoning amount of bacterial genomic data provides unparalleled opportunities for understanding the genetic and molecular bases of the degradation of organic pollutants. Aromatic compounds are among the most recalcitrant of these pollutants and lessons can be learned from the recent genomic studies of *Burkholderia xenovorans* LB400 and *Rhodococcus* sp. strain RHA1, two of the largest bacterial genomes completely sequenced to date.

These studies have helped expand our understanding of bacterial catabolism, non-catabolic physiological adaptation to organic compounds, and the evolution of large bacterial genomes. First, the metabolic pathways from phylogenetically diverse isolates are very similar with respect to overall organization. Thus, as originally noted in pseudomonads, a large number of "peripheral aromatic" pathways funnel a range of natural and xenobiotic compounds into a restricted number of "central aromatic" pathways. Nevertheless, these pathways are genetically organized in genus-specific fashions, as exemplified by the b-ketoadipate and Paa pathways.

Comparative genomic studies further reveal that some pathways are more widespread than initially thought. Thus, the Box and Paa pathways illustrate the prevalence of non-oxygenolytic ring-cleavage strategies in aerobic aromatic degradation processes. Functional genomic studies have been useful in establishing that even organisms harboring high numbers of homologous enzymes seem to contain few examples of true redundancy. For example, the multiplicity of ring-cleaving dioxygenases in certain rhodococcal isolates may be attributed to the cryptic aromatic catabolism of different terpenoids and steroids. Finally, analyses have indicated that recent genetic flux appears to have played a more significant role in the evolution of some large genomes, such as LB400's, than others.

However, the emerging trend is that the large gene repertoires of potent pollutant degraders such as LB400 and RHA1 have evolved principally through more ancient processes. That this is true in such phylogenetically diverse species is remarkable and further suggests the ancient origin of this catabolic capacity.

ANAEROBIC BIODEGRADATION OF ORGANIC POLLUTANTS

Anaerobic microbial mineralization of recalcitrant organic pollutants is of great environmental significance and involves intriguing novel biochemical reactions. In particular, hydrocarbons and halogenated compounds have long been doubted to be degradable in the absence of oxygen, but the isolation of hitherto unknown anaerobic hydrocarbon-degrading and reductively dehalogenating bacteria during the last decades provided ultimate proof for these processes in nature.

Many novel biochemical reactions were discovered enabling the respective metabolic pathways, but progress in the molecular understanding of these bacteria was rather slow, since genetic systems are not readily applicable for most of them. However, with the increasing application of genomics in the field of environmental microbiology, a new and promising perspective is now at hand to obtain molecular insights into these new metabolic properties.

Several complete genome sequences were determined during the last few years from bacteria capable of anaerobic organic pollutant degradation. The ~4.7 Mb genome of the facultative denitrifying *"Aromatoleum aromaticum"* strain EbN1 was the first to be determined for an anaerobic hydrocarbon degrader (using toluene or ethylbenzene as substrates). The genome sequence revealed about two dozen gene clusters (including several paralogs) coding for a complex catabolic network for anaerobic and aerobic degradation of aromatic compounds.

The genome sequence forms the basis for current detailed studies on regulation of pathways and enzyme structures. Further genomes of anaerobic hydrocarbon degrading bacteria were recently completed for the iron-reducing species *Geobacter metallireducens* and the perchlorate-reducing *"Dechloromonas aromatica"*, but these are not yet evaluated in formal publications. Complete genomes were also determined for bacteria capable of anaerobic degradation of halogenated hydrocarbons by halorespiration: the ~1.4 Mb genomes of *Dehalococcoides ethenogenes* strain 195 and *Dehalococcoides* sp. strain CBDB1 and the ~5.7 Mb genome of *Desulfitobacterium hafniense* strain Y51. Characteristic for all these bacteria is the presence of multiple paralogous genes for reductive dehalogenases, implicating a wider dehalogenating spectrum of the organisms than previously known. Moreover, genome sequences provided unprecedented insights into the evolution of reductive dehalogenation and differing strategies for niche adaptation.

DETECTING BACTERIA INVOLVED IN BIODEGRADATION

Traditional molecular analyses have led to great understanding of the microbial diversity in natural systems. These approaches can tell the presence of a particular group of bacteria, but does not address the activity. Molecular methods, including microautoradiography, mRNA analysis, growth assays, and incorporation of stable isotopes, can be used to determine which bacteria are involved in biodegradation of chemical pollutants. This information leads to a greater understanding of the role of microbial community structure and function with respect to bioremediation.

BIOREMEDIATION WITH EXTRACELLULAR ELECTRON TRANSFER

Geobacter species are often the predominant organisms when extracellular electron transfer is an important bioremediation process in subsurface environments. Therefore, a systems biology approach to understanding and optimizing bioremediation with *Geobacter* species has been initiated with the ultimate goal of developing *in silico* models that can predict the growth and metabolism of *Geobacter* species under a diversity of subsurface conditions. To date, these studies have included sequencing the genomes of multiple *Geobacter* species and detailed functional genomic/physiological studies on one species, *Geobacter sulfurreducens*. Genome-based models of several *Geobacter* species that are able to predict physiological responses under different environmental conditions are now available.

Quantitative analysis of gene transcript levels during *in situ* uranium bioremediation has demonstrated that it is possible to track *in situ* rates of metabolism and the *in situ* metabolic state of *Geobacter* in the subsurface. Initial attempts to link *in silico Geobacter* models with existing subsurface hydrological and geochemical models are underway. It is expected that this systems approach to bioremediation with *Geobacter* will provide the opportunity to evaluate multiple *Geobacter*-catalyzed bioremediation strategies *in silico* prior to field implementation, thus providing substantial savings when initiating large-scale *in situ* bioremediation projects for groundwater polluted with uranium and/or organic contaminants.

SIGNALLING NETWORKS AND POLLUTANT BIOSENSORS

The two elements needed for an efficient utilization of aromatic compounds by bacteria are the enzymes responsible for their degradation and the regulatory elements that control the expression of the catabolic operons to ensure the more efficient output depending on the presence/absence of the aromatic compounds or alternative environmental signals. Transcriptional regulation seems to be the more common and/or most studied mechanism of expression of catabolic clusters, although post-transcriptional control also plays an important role. Transcription is dependent on specific regulators that channel the information between specific signals and the target gene(s).

A more complex network of signals connects the metabolic and the energetic status of the cell to the expression of particular catabolic clusters, overimposing the specific transcriptional regulatory control. In general, the regulatory networks that control the operons involved in the catabolism of aromatic compounds are endowed with an extraordinary degree of plasticity and adaptability.

Elucidating such regulatory networks pave the way for a better understanding of the regulatory intricacies that control microbial biodegradation of aromatic compounds, which are key issues for the rational design of more efficient recombinant biodegraders, bacterial biosensors, and biocatalysts.

BIOAVAILABILITY, CHEMOTAXIS, AND TRANSPORT OF ORGANIC POLLUTANTS

Bacterial pathways for the degradation of organic pollutants have been the subject of intense study for decades. However, important physiological events that precede the catabolism of these compounds have recently been receiving significant scientific attention. Bioavailability, or the amount of a substance that is physiochemically accessible to microorganisms is a key factor in the efficient biodegradation of pollutants. Chemotaxis, or the directed movement of motile organisms towards or away from chemicals in the environment is an important physiological response that may contribute to effective catabolism of molecules in the environment. In addition, mechanisms for the intracellular accumulation of aromatic molecules via various transport mechanisms are also important.

SOLVENT TOLERANCE AND PUMPS THAT EXTRUDE TOXIC CHEMICALS

Organic solvents are toxic for microorganisms because they dissolve in the cytoplasmic membranes, a process that alters the membrane's physical structure and renders the cell unable to synthesize ATP. The degree of toxicity varies depending on the chemical and the strain involved, and chemical toxicity correlates with the partition coefficient of the compound in a mixture of octanol and water. Microbial tolerance to solvents can be mediated by physical and biochemical barriers.

Physical barriers are usually based on increased membrane rigidity through the alteration of the cis/trans unsaturated fatty acid ratio, the increase in the saturated:unsaturated fatty acid ratio, or alteration in the phopholipd head groups. Although these barriers counteract the effect of initial toxicity, long-term resistance is based on the active extrusion of solvents, which is mainly mediated by extrusion pumps.

Genomic analyses in Gram-negative bacteria have revealed that the resistance-nodulation-cell division (RND) family of efflux pumps is the main group involved in the removal of solvents from the cell. These pumps are made up of three components that span the membranes and extrude solvents from the inner membrane or cytoplasm to the outer medium. The level of expression of these extrusion pumps is finely modulated by transcriptional regulators belonging to different families, but which act in a similar fashion. Some regulators act as repressors that prevent access of the RNA polymerase to the promoter region of the cognate operon. These regulators recognize multiple drugs through a series of overlapping binding pockets, and upon the binding of an effector, transmit a signal to the DNA binding region so that the regulator is released and the genes encoding the efflux pumps are transcribed.

DESIGNER STRAINS FOR ENHANCED BIODEGRADATION

Using well-established tools from metabolic engineering and biochemistry, efforts have been made on engineering microbes to function as "designer biocatalysts," in which certain desirable traits are brought together with the aim of optimizing the rate and specificity of biodegradation pathways. One common bottleneck is the transport of pollutants across the cell membrane, which limits the overall rate in many microbial biodegradation. An example is for a class of neurotoxic organophospates, which are used extensively as pesticides and chemical warfare agents.

Although an enzyme, organophosphorus hydrolase (OPH), has been shown to degrade these pesticides effectively, the use of whole cell detoxification is limited by the transport barrier of substrates across the cell membrane. Display of OPH onto the cell surface has been employed to bypass this transport barrier, resulting in 7-fold faster degradation compared to whole cells expressing OPH intracellularly. This simple approach typified the unique combination of chemical engineering principle with modern genetics, and has been similarly employed for other useful environmental applications such as the display of metal-binding proteins described earlier.

Although only fairly simple enzymes or peptides are successfully displayed so far, continued development in this area should pave the way for the successful display of more complex enzymes, such as dioxygenases or monooxygenases, enabling a broader class of pollutants to be targeted. Recruiting different pathways into a designer microbe is another powerful approach to enhance biodegradation. Very often, these pathways are combined with other existing pathways to enable complete biodegradation. For example, construction of a hybrid strain which is capable of mineralizing components of a benzene, toluene, and *p*-xylene mixture simultaneously was attempted by redesigning the metabolic pathway of *Pseudomonas putida*.

A hybrid strain carrying both the *tod* and the *tol* pathways was constructed and was found to mineralize a benzene, toluene, and *p*-xylene mixture without accumulation of any metabolic intermediate. Since the number of known biodegradation pathways is increasing everyday, this in combination with the increasing number of genome sequence elucidated for environmental microbes, should allow us to rationally combine useful pathways across species into any desirable combinations using tools available from metabolic engineering. In this respect, it will also be interesting to see whether multiple enzymes can also be displayed onto the surface, allowing sequential degradation to occur without any uptake limitation. The challenges here are to devise strategies that will allow not only multiple enzyme display, but also the display of complex enzymes without compromising integrity and viability.

The major bottlenecks for *ex-situ* or *in-situ* applications are capitalized. Another promising approach for success remediation is the introduction of biodegradation pathways into microbes that thrive in the contaminated environment. *Deinococcus radiodurans* is a soil bacterium that can survive acute exposures to ionizing radiation of 15,000 Gy without lethality. A recombinant *D. radiodurans* strain expressing toluene dioxygenase was shown to effectively oxidize toluene, chlorobenzene, and TCE in a highly irradiating environment. The recombinant strains were also tolerant to the solvent effects of toluene and TCE at levels exceeding those of many radioactive waste sites. The prospect of using this strategy to alleviate the toxicity of radionuclides and heavy metals, and to provide efficient treatment for a variety of organic wastes is very promising. Similarly selective advantages can be achieved by exploiting the synergistic plant-microbe relationship in a rhizoshere.

This strategy was recently reported using a wheat rhizosphere system for the detoxification of soil-borne trichloroethylene (TCE). The toluene *o*-monooxygenase (Tom) gene was introduced into *Pseudomonas fluorescents,* a bacterium that colonizes the wheat root, enabling the establishment of a bacterium- plant-soil microcosm. Treatment of TCE-contaminated surface and near-surface soil was demonstrated, with more than 63 per cent of the initial TCE removed within 4 days. The most attractive aspect of this technology is the low cost associated since only expenses required for planting is necessary. Our group is actively pursuing this strategy by engineering both metal-binding and TCE capabilities into a single rhizobacterium, allowing it to retain TCE degradation in the presence of high level of metal contamination.

Since over 40 per cent of all superfund sites in the U.S. are co-contaminated with organic pollutants and heavy metals, the use of plant-microbe rhizoremediation will provide an ecologically sound and safe method for restoration and remediation. This will also represent an excellent opportunity for chemical engineers working primarily with microorganisms to collaborate with others focusing on plant, combining the unique features of rhizoremediation with phytoremediation.

A notable opportunity that has been so far overlooked by most chemical engineers is the production of valuable products and energy directly from wastewater. Of particular interest is the possibility of biohydrogen and bioelectricity production. In most cases, only natural microorganisms are exploited, resulting in fairly modest yields. However, this poor conversion also represents an excellent opportunity to employ the tools from metabolic engineering, protein engineering, molecular evolution, and system biology for the discovery of novel microorganisms with significantly improved efficiencies.

ENZYME ENGINEERING FOR IMPROVED BIODEGRADATION

The ever-increasing information regarding the structure and function of enzymes and pathways involved in biodegradation of recalcitrant pollutants offers opportunities for improving enzymes or entire pathways by genetic engineering. Control mechanism and enzyme properties can be tailored by site directed mutagenesis, which is often guided by computerassisted modeling of the three-dimensional(3-D) protein structures. For example, site directed approaches have been applied to enlarge the binding pocket of haloalkane dehalogenase, resulting in several-fold faster dechlorination of dichlorohexane.

However, no mutant tested could utilize the more bulky substrates, such as TCE, suggesting limitations using this structural based approach. Perhaps the use of computational methods to predict subtle and distal changes in the protein backbone without perturbing the overall protein structure could be used to further improve enzyme function and stability. Site-directed or rational approaches can often fail because it is known that mutations far from the active site can modulate catalytic activity or substrate recognition but are difficult to predict *a priori.* These methods are also restrictive because they allow the exploration of only a very limited sequence space at a time. This is clearly indicated by the creation of several chimeric enzymes guided by sequence comparison between two similar biphenyl dioxygenases.

Although the resulting variants were capable of hydroxylating both double *ortho-* and *para-*substituted PCBs, combining the substrate range of the two parental enzymes, no new activity was observed. In this case, irrational approaches such as DNA shuffling, which allow the cross-breeding of genes between diverse classes of species, can be a preferable alternative to direct the evolution of enzymes or pathways with highly specialized traits. In two independent studies, the substrate range of biphenyl dioxygenases toward PCBs has been successfully extended using directed evolution.

Variants were obtained by random shuffling of DNA segments between the large subunit of two wild type biphenyl dixoxygenases. Several variants had extended substrate ranges for PCBs exceeding those of the two parental enzymes. Similar attempts to extend the substrate specificity of toluene *ortho-*monooxygenase (TOM) and OPH have been successful. In both cases, DNA shuffling was combined with simple plate screening assays, resulting in rapidly degradation of virtually nondegradable substrates. These examples are perhaps the best reminder, suggesting that other important biodegradation enzymes could be similarly improved with this strategy since the number of related dioxygenases, monoxygenases, and hydrolases for different pollutants are virtually unlimited. Molecular evolution is probably the most useful way for evolving biodegradation enzymes for extended substrate specificities since microbial degradation of xenobiotics is usually by cometabolism and does not exert a natural selective pressure on bacteria. Computational methods that are useful to guide experimental design for directed evolution may be used to predict the optimal number of mutants that must be screened. Moreover, an optimal design of the parental DNA sequence set will allow a more focused probing of sequence space in only those regions that are likely to yield functional hybrids and should lead to a more efficient utilization of experimental resources, saving time and effort by reducing the number of evolutionary cycles.

EVOLUTIONARY AND GENOMIC APPROACHES TO BIODEGRADATION

Evolutionary approaches are extremely useful for optimization of an entire biodegradation pathway comparing to stepby- step modifications offered by rational design. This was recently demonstrated by the modification of an arsenic resistance operon using DNA shuffling. Cells expressing the optimized operon grew in up to 0.5 M arsenate, a 40-fold increase in resistance. Moreover, a 12-fold increase in the activity of one of the gene products (arsC) was observed in the absence of any physical modification to the gene itself. The authors speculate that modifications to other genes in the operon effect the function of the *arsC* gene product.

Such unexpected but exciting results are more likely to be realized using irrational approaches. This strategy is particularly attractive since the ultimate goal of many remediation approaches is for complete mineralization of the pollutants, and the concurrent optimization of an entire pathway will allow the efficient search for the correct coordination between a complex set of biodegradation reactions.

Along the same line, recent advances in genome shuffling between species, which allow the exchange and recombination of diverse pathways into a single species, will further accelerate the discovery of novel microbes that are useful for the remediation of even a complex mixture of pollutants. The availability of bacterial genomes relevant to biodegradation in recent years has allowed the feasibility to study the complex interactions between cellular reactions from a genomic and proteomic level. A quantitative understanding of how cells function requires every gene and protein to be placed in their dynamic context, which entails the integrated consideration of many interacting components.

From this perspective, a system biology approach is necessary to predict the functioning of an organism in a complex environment and to describe the outcome of the thousands of individual reactions that are simultaneously taking place in a microbial cell. So far, such prokaryotic models have been limited primarily to *E. coli* and a few pathogens. However, similar modeling approaches should be able to predict contaminant bioremediation by microorganisms that are known to predominate in polluted environments. Recently, de Lorenzo and coworkers presented a pioneering study on the characteristics of the "global biodegradation network", in which they considered the global pool of known chemical reactions implicated in biodegradation regardless of their microbial hosts. The characteristics of this network support an evolutionary scenario in which the reactions evolved from the central metabolism toward more diversified reactions, allowing us to understand the evolution of new pathways for the degradation of xenobiotics and provide the basis for predicting the abilities of chemicals to undergo biological degradation, and for quantifying the evolutionary rate for their elimination in the future.

This type of analysis, when coupled with the predictive approach for microbial catabolism using the University of Minnesota Biocatalysis/ Biodegradation Database (UM-BBD) as a knowledge base and various sets of heuristic rules, will lead to untapped and improved strategies for bioremediation. This represents an excellent opportunity for chemical engineers who are already involved with system biology, and will undoubtedly evolve into an important research direction within the next 5 years.

PROCESS ENGINEERING FOR IMPROVED BIODEGRADATION

Scientists have generally considered biological treatment processes too inefficient to challenge chemical treatment processes, particularly in treating large volumes of waste. The typically long contact time of 10 s or even longer between the pollutants and microbes would require an impractically large process for a practical treatment plant. However, it is well known that biological processes are generally safer, greener, and cheaper to run. Recently, advances in process design have brought this dream into reality by converting a chemical scrubber that removes hydrogen sulfide (a gas that smells like rotten-egg) into a biotrickling filter, which reduces hydrogen sulfide into non-smelly sulfate.

To achieve a gas contact time that is comparable with that of a chemical scrubber, the researchers passed contaminated air through the biofilter at a high velocity using a packing material with a high surface area. The unusually high air velocity resulted in a higher gas-film mass-transfer coefficient and outstanding H2S removal. After almost 2 years of continuous operation, the bioscrubbers still retained H2S removal efficiencies of more than 98 per cent. This finding is important as it suggests that substantial improvement in process performance could be achieved using a very simple design improvement, based on chemical engineering principles.

The obvious challenges are whether similar conversions of chemical treatment processes into biological processes could be achieved, offering the same treatment capacity that are much cheaper and safer. Considering the cost benefit and environmental impact of switching to biological treatment processes, this is one opportunity that chemical engineers cannot afford to miss. Undoubtedly, the numbers of problems that define environmental biotechnology are not restricted. However, it is our opinion that these perhaps represent the best opportunities for chemical engineers to make significant contributions, because of the unique intertwining of molecular biology, microbiology, reaction engineering, transport phenomena, and process design.

Because of the interdisciplinary nature of this research area, a successful chemical engineer in the field must receive training in an interdisciplinary environment where expertise from different areas is available, a trend that is consistent with the evolving nature of chemical engineering education. Although the ability to predictively design microbes or enzymes for any given remediation remains an overwhelming task, the increasing understanding of fundamental mechanistic principles generated from both genomic research or directed evolution will likely lead to the emergence of novel solutions for improved bioremediation.

Chemical engineers must embrace this important opportunity in a fashion similar to the efforts invested on the early evolution of metabolic engineering and tissue engineering, two important areas that are now partially defined by chemical engineers.

MICROBIAL BIODEGRADATION, BIOREMEDIATION AND BIOTRANSFORMATION

The use of living organisms for the recovery/ cleaning up of a contaminated medium (soil, sediment, air, water). The process of bioremediation might involve introduction of new organisms to a site, or adjustment of environmental conditions to enhance degradation rates of indigenous fauna.

Bioremediation can be applied to recover brownfields for development and for preparing contaminated industrial effluents prior to discharge into waterways. Bioremediation technologies are also applied to contaminated wastewater, ground or surface waters, soils, sediments and air where there has been either accidental or intentional release of pollutants or chemicals that pose a risk to human, animal or ecosystem health.

Different approaches to bioremediation take advantage of the metabolic processes of different organisms for degradation, or sequestering and concentration, of different contaminants. For example, soil bioremediation might be performed under either aerobic or anaerobic conditions, and involve optimization of the metabolic pathways of bacteria or fungi for degradation of hydrocarbons, aromatic compounds or chlorinated pesticides. Phytoremediation is a bioremediation process using plants and is often proposed for bioaccumulation of metals.

Bioremediation using genetically engineered microorganisms (GEMs), carrying recombinant proteins, is still relatively uncommon due to regulatory constraints related to their release and control. Other methods of enzyme optimization that do not include gene cloning technqiues, might be applied to indigenous microorganisms in order to enhance their pre-existing traits.

Examples: Nutrients were added to the soil to enhance bacterial degradation of contaminants and increase the rate of bioremediation on the brownfield site.

Interest in the microbial biodegradation of pollutants has intensified in recent years as mankind strives to find sustainable ways to cleanup contaminated environments.These bioremediation and biotransfo-rmation methods endeavour to harness the astonishing, naturally occurring, microbial catabolic diversity to degrade, transform or accumulate a huge range of compounds including hydrocarbons (*e.g.* oil), polychlorinated biphenyls (PCBs), polyaromatic hydrocarbons (PAHs), pharmaceutical substances, radionuclides and metals.

Major methodological breakthroughs in recent years have enabled detailed genomic, metagenomic, proteomic, bioinformatic and other high-throughput analyses of environmentally relevant microorganisms providing unprecedented insights into key biodegradative pathways and the ability of organisms to adapt to changing environmental conditions.

The elimination of a wide range of pollutants and wastes from the environment is an absolute requirement to promote a sustainable development of our society with low environmental impact. Biological processes play a major role in the removal of contaminants and they take advantage of the astonishing catabolic versatility of microorganisms to degrade/convert such compounds.

New methodological breakthroughs in sequencing, genomics, proteomics, bioinformatics and imaging are producing vast amounts of information. In the field of Environmental Microbiology, genome-based global studies open a new era providing unprecedented *in silico* views of metabolic and regulatory networks, as well as clues to the evolution of degradation pathways and to the molecular adaptation strategies to changing environmental conditions.

Functional genomic and metagenomic approaches are increasing our understanding of the relative importance of different pathways and regulatory networks to carbon flux in particular environments and for particular compounds and they will certainly accelerate the development of bioremediation technologies and biotransformation processes.

6

Chemical Engineering Perspective of Environmental Biotechnology

Environmental biotechnology refers to the utilization of microorganisms to improve environmental quality. Although the field of environmental biotechnology has been around for decades, starting with the use of activated sludge and anaerobic digestion in the early 20th century by civil engineers, the introduction of new technologies from modern microbiology and molecular biology has enabled engineers and scientists to tackle the more contemporary environmental problems such as detoxification of hazardous chemicals.

Chemical engineers are uniquely poised to contribute in this emerging area since many of the potential solutions require a combined perspective from modern biology and process engineering, two areas where chemical engineers excel. For example, the realization of environmental biotechnology into practical solutions requires the implementation of process design, which is the foundation of the chemical engineering discipline. The same is also true at the cellular level, where the functions of cells are determined primarily by networks of specific catalytic reactions.

The nature and activities of these networks are dictated by the genetic information, thereby defining the ways in which engineers can influence cellular functions and metabolic capabilities toward the designed of improved biocatalysts for environmental remediation. Although, superficially these strategies seem distant from traditional chemical engineering, a deeper inspection reveals that the design of biological catalysts, based on defined techniques from biochemistry and biology is indeed parallel to our understanding of chemical kinetics, transport, separation, and control.

In addition, advances in genomics and proteomics are providing opportunities to predict, in a quantitative manner, the potential manipulations necessary. Even though chemical engineers are well prepared to contribute new research directions in environmental biotechnology, only a few are working in this area today. Fortunately, the number of chemical engineers showing interest is growing every year and with the recent research emphasis on biotechnology, it is easy to envision that many others will join this exciting research area in the near future. We will attempt to highlight opportunities available for chemical engineers to make significant contributions and their future challenges.

EVALUATION IN ENVIRONMENTAL ENGINEERING

PRECIPITATIVE SOFTENING

Precipitative softening is the most commonly used process for removing hardness cations in drinking water treatment plants. It is based on the precipitation of calcium as calcium carbonate and magnesium as magnesium hydroxide.

$$Ca^{+2} + CO_3^{-2} \leftrightarrow CaCO_3$$
$$Mg^{+2} + 2OH^- \leftrightarrow Mg(OH)_2 \downarrow$$

Chemical Basis

For this purpose, lime $Ca(OH)_2$ and soda ash Na_2CO_3 are added in proportion to the hardness and alkalinity. This is called lime-soda softening. In order to precipitate calcium, the solubility product constant for calcium carbonate must be exceeded. For magnesium removal, the solubility product constant for the hydroxide must be exceeded.

$$[Ca^{+2}][CO_3^{-2}] = 10^{-8.15}$$
$$[Mg^{+2}][OH]_2 = 10^{-9.2}$$

Since at the pH of most natural waters, hydroxide and carbonate concentrations are quite low (bicarbonate concentration may be high, however), high concentrations of calcium and magnesium can exist in solution and the concentration products will still be below their respective solubility product constants. In other words, the magnesium and calcium will not precipitate. If, either the carbonate or hydroxide concentrations can be made to increase, precipitation will occur. By adding lime, naturally-present bicarbonate is converted to carbonate. In order for this reaction to occur at a reasonable rate, enough lime must be added to completely react with the bicarbonate present. If manganese is also to be removed, then additional lime must be added to account for the precipitation of manganese hydroxide. Experience has shown that more must be added to achieve a sufficiently high pH for rapid precipitation of magnesium. Note that calcium removal results in removal of bicarbonate and carbonate as well.

$$2\ HCO_3 + Ca(OH)_2 \rightarrow 2\ CO_3^{-2} + Ca^{+2} + 2H_2O$$
$$Mg^{+2} + Ca(OH)_2 \rightarrow Mg(OH)_2 \downarrow + Ca^{+2}$$

Let us restate the lime softening chemistry. Once all of the bicarbonate is converted to carbonate, addition of lime results in an accumulation of hydroxide and a rise in pH. This excess hydroxide will eventually result in the precipitation of magnesium hydroxide. Therefore, for magnesium removal to approach completion, the required molar dose of lime is equal to one half the bicarbonate concentration plus the magnesium concentration. In practice, it is necessary to raise the pH to about 11 to accelerate manganese precipitation. This requires the addition of an excess of lime amounting to one-half mM/L.

$$[\text{Lime Dose}] = 0.0005 + [Mg^{+2}] + 0.5\,[HCO_3]$$

where magnesium, bicarbonate and lime are all in moles/liter.

When the amount of bicarbonate originally-present is insufficient to precipitate all the calcium (originally-present and added), soda ash must be added. The amount of soda ash needed will be equal to the "bicarbonate deficit". This can be calculated from the amount of lime added plus the calcium originally-present $[Ca^{+2}]$ minus the bicarbonate originally present $[HCO_3]$. This is represented in equation. If the calculated molar dose is less than zero, the system is not bicarbonate limited and soda ash should not be added. In such a case precipitation of carbonate is incomplete, while calcium and magnesium precipitation approach completion.

$$[\text{Soda Ash Dose}] = 0.0005 + [Mg^{+2}] + [Ca^{+2}]-0.5[HCO_3]$$

Post-precipitation

Precipitative softening is slow chemical process. The precipitation of calcium is a good example of a chemical equilibrium which is kinetically controlled within the time frame of interest (minutes-hours).

For this reason, precipitation of hardness cations will continue into the distribution system unless something is done to stop it. Commonly, the pH is lowered by recarbonation (addition of carbon dioxide) or acidification prior to discharge into the distribution system. The desired final pH may be calculated by using the solubility product for calcium carbonate and the acid dissociation of bicarbonate.

For example, rearranging equation, we get:

$$[CO_3^{-2}] = 10^{-8.15}/[Ca^{+2}]$$

Similarly the equilibrium equation for the dissociation of bicarbonate to carbonate can be rearranged to give:

$$[H^+] = 10^{-10.33}[HCO_3^-]/[CO_3^{-2}]$$

Now substituting equation gives:

$$[H^+] = 10^{-2.18}[Ca^{+2}][HCO_3^-]$$

and taking the negative of the logarithm of both sides:

$$pHs = 2.18\text{-}\log[Ca^{+2}]\text{-}\log[HCO_3^-]$$

We use the subscript "s" in equation to indicate that this is the pH of saturation with respect to calcium carbonate. Above this pH the system is no longer in equilibrium (the solubility product constant is exceeded) and calcium carbonate should be precipitating.

Below this pH, the system is undersaturated with respect to calcium carbonate. In this case calcium carbonate will not precipitate, and in fact, any precipitated calcium carbonate that comes in contact with such a water will tend to dissolve.

LANGELIER INDEX

The tendency of a water to precipitate or dissolve calcium carbonate can be characterized numerically by the Langelier Index. This value is defined as the difference between the actual pH (pHa) and the saturation pH.

$$\text{L. I.} = \text{pHa} - \text{pHs}$$

If,

L. I. > 0, system is oversaturated

L. I. < 0, system is undersaturated

The desired Langelier Index is usually not zero, but rather some small positive number on the order of +0.8 to +1.0. This is due to the tendency of a precipitating water to form a protective coating of calcium carbonate around the inside of iron pipes. This coating will often inhibit corrosion in such systems. However, too much precipitation will lead to clogging of the pipes and reduced hydraulic capacity.

Ionic Strength Effects

The equilibrium constants used to calculate pHs are all based on the ideal case of infinite dilution (m=0). In dilute waters, this is a good approximation, however, waters subject to softening generally have higher ionic strengths. For this reason, corrections must be made on equation. All rigorous chemical equilibria are expressed in terms of activity ($\{X^-\}$. In dilute solution, these activities are nearly equal to the analytical concentrations ($[X^-]$), and we generally express them as such. However, in waters of high ionic strength, differences between concentration and activities must be taken into account, usually by means of an activity coefficient, γ.

$$\{X^-\} = \gamma x \, [X^-]$$

When activity coefficients are inserted into the equations for the solubility product of $CaCO_3$ and the dissociation of HCO_3, and equation is re-derived, one gets:

$$\text{pHs} = 2.18 - \log_\gamma \text{Ca} - \log[Ca^{+2}] - \log_\gamma HCO_3 - \log[HCO_3]$$

Activity coefficients may be calculated from the Guntelberg approximation of the Extended Debye-Hückel formula.

$$\lig \gamma = -0.5(z)^2 \frac{\sqrt{1}}{1+\sqrt{1}}$$

The ionic strength, I, can be estimated from conductivity by the Russell approximation, or from TDS by the Langelier approximation.

$$I = 1.6 \times 10\text{-}5 \times \text{Specific Conductance (in mmho/cm)}$$

$$I = 2.5 \times 10\text{-}5 \times \text{TDS (in mg/L)}$$

COAGULATION JAR TESTING

Coagulation jar testing is a simple laboratory-scale test intended to simulate many of the processes that occur during full-scale coagulation and settling. It is especially useful for selecting optimum coagulant doses, coagulant types and combinations. A variable-speed six-paddle stirrer is used for these tests. Coagulants are added to the "jars" (actually 1000 mL or 600 mL beakers or square plexi-glass containers) during a rapid mix period. This is done with a pipet while the paddles are rotated at their highest speed. Next a period of slow mix is used to simulate flocculation. Finally the paddles are removed from the beakers and the floc is allowed to settle.

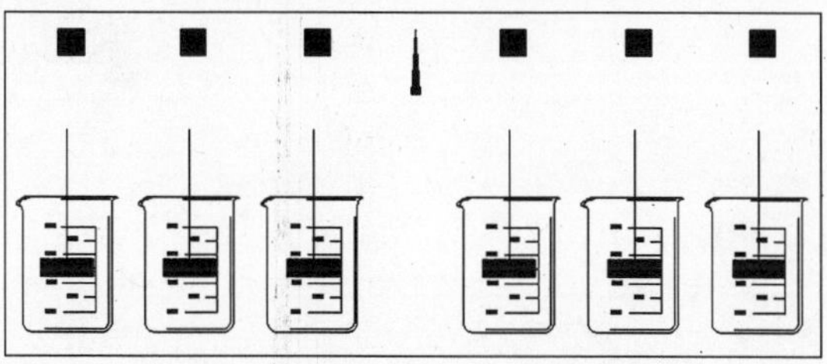

Coagulation processes are defined by the chemistry of the particular coagulants used. In the case of alum (aluminum sulfate), a zone of minimum solubility exists near pH 6. This is the region where the aluminum hydroxide precipitates most rapidly and completely.

CHLORINATION BYPRODUCT PRECURSORS

Formation Potentials vs Simulated Distribution System Tests

One can distinguish 3 categories of DBP precursor tests:

1. The high-dose formation potential (HDFP) tests;
2. The low-dose formation potential (LDFP) tests; and
3. The SDS tests.

The HDFP is characterized by two attributes:

1. It is not used to simulate any particular water system or water treatment scenario; and
2. It uses a sufficiently high chlorine dose so that the residual remains high and nearly constant with contact time, and independent of chlorine demand (many use a dose of 20 mg/L for samples with a chlorine demand of 8 mg/L or less, so that the residual is always between 12 and 20 mg/L).

The HDFP is trying to be an unbias precursor test. This means it exhibits the same per cent precursor recovery regardless of precursor concentration. Changing TOC concentrations in waters of high bromide does not present a complication with the HDFP, because the TOC "sees" the same oxidizing environment, regardless of what the actual TOC concentration is. This is not the case for the other tests.

The LDFP is also characterized by two attributes:

1. It is not used to simulate any particular water system or water treatment scenario; and
2. It uses a low chlorine dose so that the residual at the end of the contact time is close to what is commonly found in the taps of most US public water systems.

The exact chlorine dose is adapted in some way to the sample's chlorine demand. Note that the Uniform Formation Test (Summers, 1993) is a type of LDFP. Tests that fall into this group have an inherent bias towards higher precursor recoveries for more highly coloured waters. On the other hand, if carefully run, these tests can provide more accurate information for assessing the expected DBP concentrations at consumer's taps.

The third type of test, the SDS, is characterized by a single attribute: it is designed to simulate the formation in a particular system on a particular day. This test uses a site-specific chlorine dose, pH, temperature and contact time. The values chosen are either based on an existing system on a particular day; or they are based on a very specific scenario intended to simulate a postulated system. Each of these three types of tests has its own appropriate use. The HDFP is most useful for the assessment of process performance. It can tell you what level of precursor removal is being achieved. The LDFP is most appropriate for comparing finished waters from parallel and alternative treatment trains.

It is generally used when the precise disinfection scenario or distribution system characteristics are uncertain. It is not intended for assessing precursor removal across processes, just comparative precursor levels in the finished waters. The SDS is what should be used when the most accurate information about compliance and real-world concentrations are needed. The HDFP is the easiest test to run, and it is the most precise. This is because it is nearly independent of chlorine dose, so errors in dosing or excessive demands will introduce very little error. The LDFP and SDS tests are more labour-intensive, and they are prone to larger uncertainties.

In summary, the three types of precursor tests should be used as follows:

1. *HDFP*: For studies of isolated process performance; for understanding the behaviour of complex treatment systems; for extrapolating findings at one utility to systems elsewhere in the country
2. *LDFP*: For comparisons of parallel treatment trains at a single pilot plant.
3. *SDS*: For estimating whether a new or midified treatment system will achieve compliance.

Instantaneous DBPs, Terminal DBPs, and Available Precursors

The concentration of trihalomethanes or any other DBP in a sample of water taken from a process stream in a treatment plant or from a tap in a distribution system is referred to as simply the THM concentration or the instantaneous-THM concentration. This is the concentration that existed at the time of sampling. In many cases it is important to know how much effective DBP precursor organics are left in the sample. For this one typically holds the sample without quenching the residual oxidant, and measures the DBP concentration at some later date.

The results of this measurement are called the terminal concentration. The exact conditions used (*e.g.*, holding time, pH, additonal chlorine dose, temperature) will depend on the specifics of the study. The difference between the terminal concentration and the instantaneous concentration is then the available precursor content. This is the amount of DBP precursors, present at the time of sampling, that will react to form DBPs under the test conditions used.

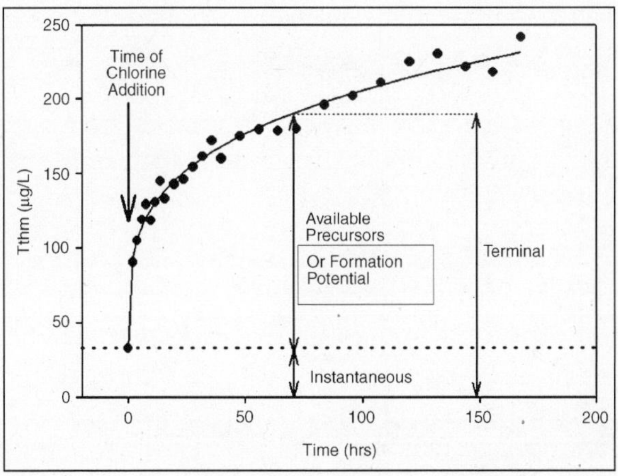

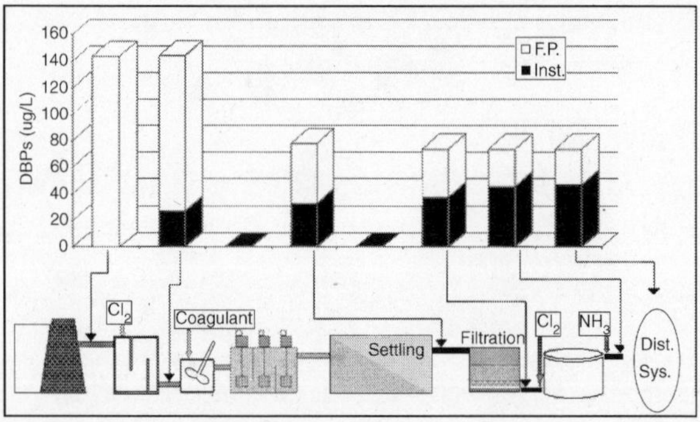

BIODEGRADABILITY TESTS

Biodegradable or assimilable organic carbon, depending on the analytical method employed, is defined as either the fraction of dissolved organic carbon (DOC) that can be used by bacteria for growth and cell maintenance or the degree to which microbial growth is stimulated by this DOC. Because of the complexity of natural organic matter, it is impossible to identify the entirety of the DOC. The characterization of biological availability of dissolved organic matter requires the either the use of bioassay techniques or possibly the use of chemical surrogates.

In heavily polluted waters, biochemical oxygen demand (BOD) or chemical oxygen demand (COD) tests are traditionally used for assessing the effectiveness of biological treatment. However, these methods are not sufficiently sensitive for use with treated drinking waters (Rittmann & Huck, 1989).

In the past decade, several methods have been proposed to determine the easily assimilable organic carbon (AOC) or the biodegradable dissolved organic carbon (BDOC) in drinking waters. Several of these methods have been review recently by Huck *et al.* (1990).

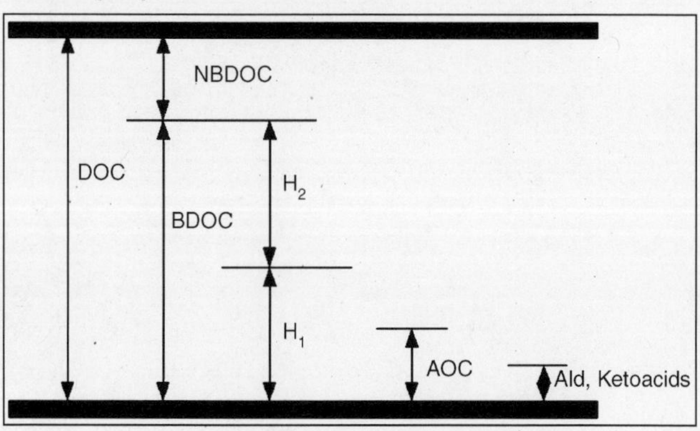

AOC Tests

The first and still the most widely-used method for measuring assimilable organic carbon was developed by van der Kooij *et al.* It requires one to follow the growth of a pure culture by a plate count on gelose. The bacteria are inoculated in the water sample, after heating to 60°C. The bacteria commonly used are Pseudomonas fluorescens P17 and Spirillum NOX. The maximum plate count generally achieved after 4 to 8 days of incubation (15°C) is related to the maximum growth of a strain on a specific substrate. It is converted into units of acetate equivalents or oxalate equivalents by referring to a calibration test on a pure acetate or oxalate standard. van der Kooij's pure culture approach suffers from several disadvantages.

Pure culture of micro-organisms has a more limited capacity for biodegration than a heterogenous bacteria population. Maximum growth counts may be influenced by the physiological state of the Pseudomonas fluorescens culture. The use of pure cultures requires special techniques and highly skilled analysts. The AOC method as it is now commonly used, may also be subject to certain types of bias and large random errors. For example, Prévost and co-workers have noted that difficulties in determining the maximum cell count can lead to significant errors, especially at high AOC concentrations. Possible growth inhibition or delayed growth caused by ozone byproducts and aluminum coagulants has also been reported. Perhaps these or other factors are responsible for some of the large anomalous increases in AOC-P17 across filtration as noted by Miltner *et al.*

Nevertheless, the pure culture approach, as exemplified by van der Kooij's, should be subject to less interlaboratory variation, because the innoculum is always the same. Some laboratories, such as the Water Research Center in England, have modified the method of van der Kooij *et al.* by using an inoculum of an autochthonous bacterial population. They have also followed the growth rate through measurements of adenosine triphosphate (ATP) concentration, rather than using plate counts. Rice and co-workers have developed another pure culture procedure that attempts to measure the ability of coliforms to grow. In this method, a pure culture of coliform (Enterobacter cloacae) is seeded and the ratio of log growth after 5 days (20°C) and time zero is determined. A higher ratio indicates a higher coliform growth response (CGR) and suggests that regrowth of coliforms may be possible in the distribution system. As with many of these alternative bioassays, the correlations with AOC and other tests are poor.

BDOC Tests

The methods of van der Kooij *et al.*, Jago and Stanfield, and Rice are based on bacterial growth. Other methods have been developed whereby the loss in organic substrate is measured. When the organic substrated is characterized by DOC measurements, the methods are termed biodegradable dissolved organic carbon (BDOC) assays. Servais and co-workers developed a bioassay that incorporates some aspects of the van der Kooij and Werner methods. The water sample containing dissolved organic matter is sterilized by membrane filtration, inoculated with a second sample of the same water and incubated for 3 weeks at 20°C.

After this time the reduction of the dissolved organic carbon (DOC) concentration is measured. In order to circumvent the long incubation times proscribed by Servais, Joret and co-workers (Joret & Levi, 1986; Joret *et al.*, 1988) proposed that water samples could be tested in the presence of an inoculum of bacteria biomass attached to sand, a more active inoculum, that could reduce the response time to 3 or 4 days.

Chemical Surrogates

A very different approach to assessing biodegradability is to use chemical surrogates for AOC or BDOC. As previously discussed, the most widely-used measure of biodegradability in the US is probably the Assimilable Organic Carbon (AOC) assay. Its chief disadvantages are its cost, labour requirements, and long analysis time, and lack of robustness. For these reasons it is not likely to be used as a routine water quality measurement, like dissolved organic carbon (DOC) or trihalomethane formation potential.

Several researchers have examined alternative chemical assays in order to avoid some of the drawbacks of AOC analysis. Reckhow *et al.* suggested that aldehydes or keto-acids could serve in this capacity. The use of aldehydes has also been supported by Zhou *et al.*and Krasner *et al.*Both groups of compounds can be measured by gas chromatography following sample extraction and chemical derivatization. Analyses are reproducible, and fast; usually taking only a few hours from start to finish.

The Keto-Acids, and probably the aldehydes are be components of the AOC-NOX, but not the AOC-P17. This is an important characteristic for a proposed surrogate, because these two strains behave very differently. It has been observed that when the two strains are added together, NOX growth is most closely aligned with the concentration of biodegradable organic matter produced by ozonation. In contrast, P17 either cannot utilize the ozone byproducts or it grows more slowly on them, so that its growth more closely reflects the level of naturally-occurring biodegradable organic compounds.

Since the precise relationship between AOC levels in finished water and problems related to bacterial regrowth in the distribution system is not known, the desirable level of AOC in drinking waters remains speculative. Many engineers have chosen as a treatment goal the reduction of post-ozonation AOC levels to their pre-treatment levels. This often means that the AOC-NOX produced by ozonation, must be removed by subsequent biological filtration. Therefore, a chemical surrogate specific for AOC-NOX could be especially useful. Studies conducted by Reckhow *et al.*, have shown that the keto-acids represent only the most biodegradable of the AOC compounds. Furthermore, there is a fraction of the AOC (even as fraction of the AOC-NOX) that is not readily degraded through water treatment. Similar observations have been made for some of the low molecular weight aldehydes.

ENGINEERING BIOSORBENTS FOR HEAVY METAL REMOVAL

Immobilization of heavy metals into biomass or precipitation through reduction to lesser bioactive metal species, such as metal sulfide are the major mechanisms employed by nature (microorganism, animals and plants) to counteract heavy metal toxicity. These natural mechanisms can be easily exploited to optimize biosorbents that are more efficient for heavy metal removal. In one example, a sulfide-dependent metal removal strategy was developed by engineering the sulfate reduction pathway into a robust bacterium *E. coli*.

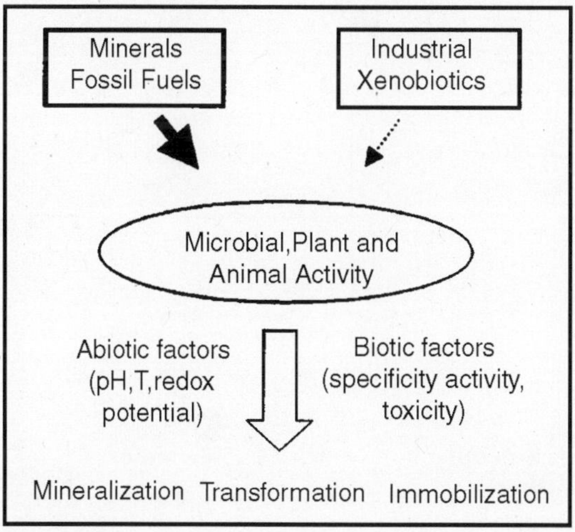

Fig. Main Sources of Environmental Pollutants and Factors Influencing their Nature Removal from the Environment.

The resulting strains produced significantly more sulfide and removed more than 98 per cent of the available cadmium under anaerobiosis. Further improvement in metal precipitation was achieved by engineering effective sulfate reduction under aerobic conditions. *E. coli* expressing both serine acetyltransferase and cysteine desulfhydrase overproduced cysteine and converted it to sulfide. The resulting strain was effective in aerobically precipitating cadmium. This aerobic approach of metal precipitation is particularly attractive as large-scale processes could be implemented under aerobic conditions. The challenges are to incorporate these genetic modifications into a robust environmental microbe that could survive and thrive under the required operation conditions. Similar success in engineering enhanced biosorbents has been achieved by displaying metal-binding peptides onto the cell surface. One example was recently reported by creating a repetitive metal-binding motif consisting of (Glu-Cys)$_n$Gly. These peptides emulate the structure of phytochelatins, metalchelating molecules that play a major role in metal detoxification in plants and fungi.

The phytochelatin analogs were presented on the bacterial surface, enhancing Cd^{2+} and Hg^{2+} bioaccumulation by 12-fold and 20-fold, respectively. Unlike nature metal-binding peptides, these *"de novo"* designed metal- binding peptides are attractive as they offer the potential of improved affinity and selectivity for heavy metals. To this end, the use of molecular modeling and evolutionary strategies may enable the rapid discovery of novel peptide sequences that are superior metal chelators. In addition to peptides, metalloregulatory proteins are another group of useful metal-binding moiety with striking affinity and specificity.

The highly specific nature of these proteins is the result of a cleverly designed genetic circuit that is tightly under their control. Examples are MerR and ArsR, which are regulatory proteins used for controlling the expression of enzymes responsible for mercury and arsenic detoxification, respectively. The high affinity and selectivity of MerR toward mercury has been exploited for the construction of microbial biosorbents specific for mercury removal.

Presence of surface- exposed MerR on an engineered strain enabled 6-fold higher Hg^{2+} biosorption. Hg^{2+} binding via MerR was very specific with no observable decline even in the presence of 100-fold excess $Cd^{2+}$and Zn^{2+} Similarly, cells overexpressing ArsR accumulated 5- and 60-fold higher levels of arsenate and arsenite with no observable binding to other heavy metals. As nature provides a wide range of metalloregulatory proteins for other important metal pollutants, this strategy may be used as a universal approach to enable selective binding of target metals of interest.

SERVING THE FUTURE OF ENVIRONMENTAL BIOTECHNOLOGY

Like white biotechnology, environmental biotechnology, often referred to as "grey biotechnology", also focuses on sustainability. For instance, environmental biotechnology deals with the treatment of sewage water, the purification of exhaust gas or the decontamination of soils or ground water using specific microorganisms. The use of organisms for the removal of contamination or pollutants is generally referred to as bioremediation. Originally, bioremediation was mainly used in cleanup operations, including the decomposition of spilt oil or slagheaps containing radioactive waste. In addition, bioremediation is also the method of choice when solvents, plastics or heavy metals and toxic substances like DDT, dioxins or TNT need to be removed. Bioadsorption processes using newly developed bioadsorbers made from renewable materials are currently being developed. These adsorbers function as ion exchangers and are used in the elimination and disposal of toxic heavy metals. The industrial use of mineral resources leads to the drastic accumulation of these pollutants in the biosphere.

The new bioadsorbers are used for the elimination of heavy metals and radionuklids from industrial wastewater, ore mine wastewater, seepage water from dumpsites or wastewater from nuclear power stations. End-of-pipe technologies and PIUS BT Biotechnological methods in environmental protection are almost exclusively typical "end-of-pipe processes" (wastewater purification, biofilters, biowashers, biogas, etc.). Nevertheless, the use of biotechnology in environmental protection not only concerns the removal of contamination generated during production, *i.e.* the degradation of harmful substances.

Modern environmental protection starts with the prevention of harmful substances prior to and during industrial production processes – production-integrated environmental protection using biotechnology. The state of Baden-Württemberg supports environ-mental biotechnology through the PIUS-BT project "Implementation of PIUS-BT in different companies and industries in Baden-Württemberg". This project was set up in Spring 2004 by the Ministry for the Environment and it was planned to run until the end of 2005. Ten companies agreed to undergo a "PIUS-BT company check" in which the companies' production processes were thoroughly checked as part of academic theses and practical term work.

It was assessed whether, and to what extent and at which point in the production process, it was possible to integrate biotechnological processes. Driving force for integrated environmental protection Biotechnology is regarded as a driving force for integrated environmental protection. The use of genetically modified enzymes in the production of textiles, food, washing agents and drugs helps to save energy and raw materials. In addition, unwanted by-products can also be prevented. It is in this way that white biotechnology – the industrial production of products using biotechnological methods – is able to make a considerable contribution to the reduction of costs and waste as well as improving the quality of the end product. That's where the interdisciplinary approach comes into its own.

A survey carried out in 2003 by the Fraunhofer Institute for Systems Technology in Karlsruhe, Germany, identified 34 product-integrated environmental biotechnological methods that are used in Baden-Württemberg. There is no doubt that biotechnology has a great potential to be an ecologically beneficial and at the same time economically profitable technology in many areas, in particular in pharmacology. In addition, white biotechnology is used in the food. Metabolic design Biotechnological and genetic engineering methods are also able to reduce the environmental burden in the field of renewable raw interesting products materials ("metabolic design"). Bioplastics are certainly among the most. The material, in the simplest case made from starch, sugar or cellulose, can be degraded biologically. It consists of compounds that can be recognised and digested by bacterial enzymes. If bioplastics are dumped on garbage dumps it takes no more than a couple of weeks for the material to vanish.

FRESH AIR IN COMPOST PLANTS

The measurement of oxygen concentrations under extreme climatic conditions, such as are present in large municipal composting pits, present particular challenges for oxygen measurement systems since they need to function reliably and independently from fluctuations in air humidity and temperature. The company J. Dittrich Elektronic GmbH and Co. KG from Baden-Baden has developed a new oxygen measurement system that is adapted to such specific composting conditions. The word "biotechnology" is very often associated with the technologies and products of an industrial sector that has made enormous progress in the last few decades.

Nevertheless, biotechnology has a very long history, which started with beer brewing, tanning or even indeed with the process of composting. Already the Romans used the oldest recycling method of the world and used soil organisms to produce effective compost soil from organic waste. Although the biotechnological method of composting has a long tradition, it is still subject to continuous improvement. In the meantime, sophisticated measurement and ventilation systems, covers or containers have been developed enabling the smooth degradation of biological waste.

Correct ventilation to prevent fermentation gas production Ensuring correct ventilation is extremely important for the operators of industrial composting plants, as the aerobic degradation of organic waste through soil bacteria depends on sufficient oxygen supply. In the presence of too low amounts of oxygen, the bacteria will turn to anaerobic degradation of the waste, which leads to foul-smelling and dangerous fermentation gases such as mercaptan, hydrogen sulphide and ammoniac. To prevent this happening, the operators need to aerate the enormous compost pits. However, care must be taken not to use too much oxygen, because this will desiccate the degrading material, which in turn will lower the microbial degradation rate. A reliable and easy to operate measurement system is required to properly adjust ventilation.

An additional problem arises from the chemically, biologically and climatically aggressive environment of the composting plants. During the intensive rotting process, relative air humidity in a compost heap can reach temporary peaks of nearly 100 percent with temperatures of over 75 °C. In order to guarantee the reliable measurement of oxygen concentration, the oxygen measurement system must not be prone to fluctuations in temperature and humidity. Electrochemical sensors need to be adapted in a complicated way that often makes them very accident-sensitive. Innovation in measurement technology – the dynamic oxygen sensor J. Dittrich Elektronic GmbH and Co. KG based in Baden-Baden (Black Forest) have developed an oxygen measurement system with a special core part: A dynamic zirconium oxide-based system that functions perfectly, and that is not affected by air humidity and temperature. It is therefore well suited to determining the oxygen concentration under extreme climatic conditions. The oxygen measurement system, marketed as MF420-0-M, has several additional advantages over electrochemical measurement devices. As the dynamic oxygen sensor does not consume chemical substances, it has a much longer lifetime than its electrochemical equivalent.

The measurement system does not determine the oxygen concentration in the same way as potentiostatic sensors, which give a relative measurement size (volume percent), but as absolute size, namely as the partial pressure of oxygen in the gas mixture. Thus, the sensor can be calibrated automatically on atmospheric air without requiring an additional reference gas, and therefore offers a simple operation.

7

Microbiology and Pollution

MICROBES

A microbe is any living thing that spends its life at a size visible sometimes only with a microscope. It is too tiny to be seen with the naked eye. Microbes are the oldest form of life on Earth. Some types have existed for billions of years. They may live as individuals or cluster together in communities. Microbes live in the water you drink, the food you eat and the air you breathe. Right now, billions of microbes are swimming in your belly and mouth and crawling on your skin! Don't worry, over 95 per cent of microbes are good for you. Microbes include bacteria, viruses, fungi, algae and protozoa. These single-cell organisms are invisible to the eye, but they can be seen with microscopes.

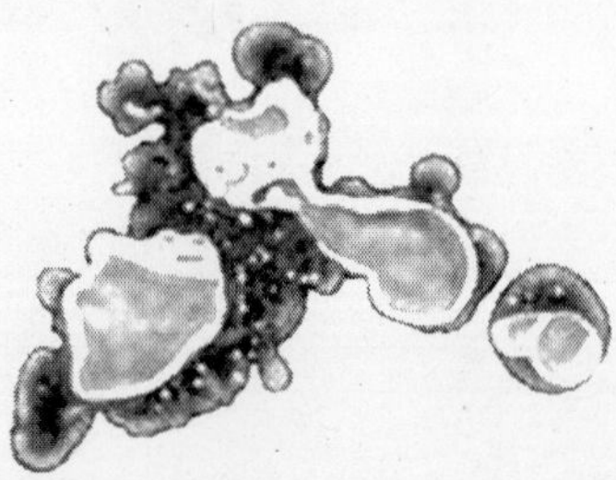

Fig. Microbe

The term microbe is short for microorganism, which means small organism. To help people understand the different types of microbes, they are grouped or classified in various ways. Microbes are very diverse and represent all the great kingdoms of life. In fact, in terms of numbers, most of the diversity of life on Earth is represented by microbes.

CLASSIFICATION OF MICROBES

The term microbe is short for microorganism, which means small organism. To help people understand the different types of microbes, they are grouped or classified in various ways. Microbes are very diverse and represent all the great kingdoms of life. In fact, in terms of numbers, most of the diversity of life on Earth is represented by microbes.

Here is an outline of the major groups of micro-organisms:

- Viruses
- Bacteria
- Algae
- Fungi
- Protozoa

Viruses

A virus is too small to be seen without a microscope. A virus is basically a tiny bundle of genetic material carried in a shell called the viral coat. Some viruses have an additional layer around this coat called an envelope. That's basically all there is to viruses. There are thousands of different viruses that come in many shapes. Many are multi-sided or polyhedral. If you've ever looked closely at a cut gem, like the diamond in an engagement ring, you've seen an example of a polyhedral shape. Unlike the diamond in a ring, however, a virus does not taper to a point, but is shaped the same all around. Other viruses are shaped like spiky ovals or bricks with rounded corners. Some are like skinny sticks while others look like pieces of looped string. Some are more complex and shaped like little spaceship landing pods.

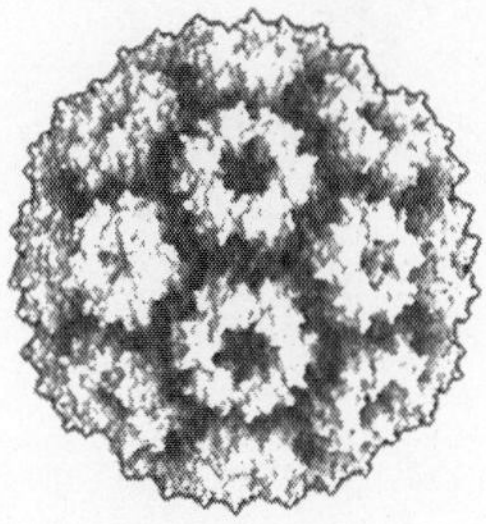

Fig. Viruses

Viruses are found on or in just about every material and environment on Earth from soil to water to air. They're basically found anywhere there are cells to infect. Viruses can infect every living thing. However, viruses tend to be somewhat picky about what type of cells they infect. Plant viruses are not equipped to infect animal cells, for example, though a certain plant virus could infect a number of related plants. Sometimes, a virus may infect one animal and do no harm, but cause a great deal of damage when it gets into a different but closely related animal.

Viruses exist to reproduce only. To do that, they have to take over suitable host cells. Upon landing on a suitable host cell, a virus gets its genes inside the cell either by tricking the host cell to pull it inside, or by connecting its viral coat with the host cell wall or membrane and releasing its genes inside, or by injecting their genes into the host cell's DNA. The viral genes are then copied many times, using the process the host cell would normally use to reproduce its own DNA. The new viral genes then come together and assemble into whole new viruses. The new viruses are either released from the host cell without destroying the cell or eventually build up to a large enough number that they burst the host cell.

Bacteria

Bacteria consist of only one cell, but they're a very complex group of living things. Some bacteria can live in temperatures above the boiling point and in cold below the freezing point. There are thousands of species of bacteria, but all of them are basically one of three different shapes. Some are rod- or stick-shaped; others are shaped like little balls. Others still are helical or spiral in shape. Some bacteria cells exist as individuals while others cluster together to form pairs, chains, squares or other groupings. Some bacteria can make their own food from sunlight, just like plants. Also like plants, they give off oxygen. Other bacteria absorb food from the material they live on or in. Some of these bacteria can live off iron or sulfur! The bacteria that live in your stomach absorb nutrients from the digested food you've eaten.

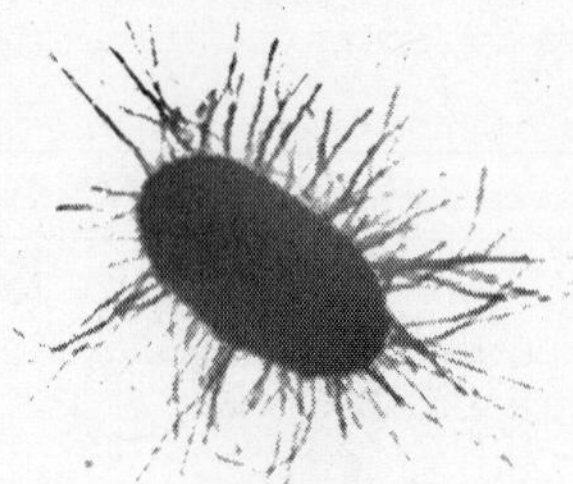

Fig. Bacteria

Some bacteria move about their environment by means of long, whip-like structures called flagella. They rotate their flagella like tiny outboard motors to propel themselves through liquid environments. They may also reverse the direction in which their flagella rotate so that they tumble about in one place. Other bacteria secrete a slime layer and ooze over surfaces like slugs. Others stay almost in the same spot. Bacteria live on or in just about every material and environment on Earth from soil to water to air and from your body to the Arctic ice to the Sahara deserts. Each square centimeter of your skin averages about 100,000 bacteria. A single teaspoon of soil contains more than a billion (1,000,000,000) bacteria.

Algae

Algae are found in fresh and salt water around the world. They can also grow on rocks and trees and in soil when enough water is available. Most algae are able to make energy from sunlight, like plants do. They produce a large amount of the oxygen we breathe. However, at some stages of their lives, some algae get their nutrients from other living things.

Diatoms are one kind of algae. They have hard shells made out of glass. When they die, these shells sink to the bottom of their watery environments. We mine deposits of these glass shells that formed hundreds of thousands of years ago to make abrasives, shiny road paint and grit in toothpaste. Diatoms come in all sorts of shapes—some are round and others are oval. Some look like leaves and others like fat commas.

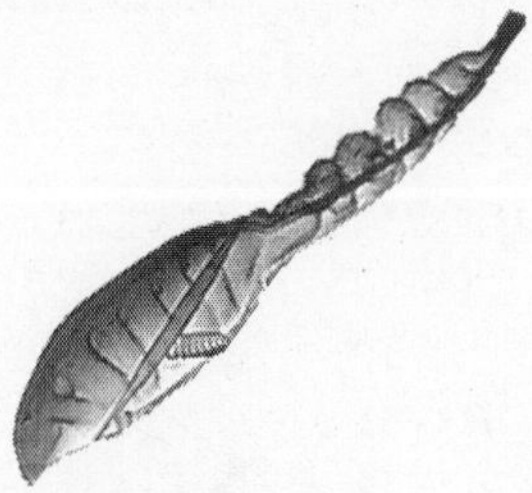

Fig. Algae

Because algae make so much oxygen, these microbes are very helpful. But sometimes certain kinds of algae can also grow in such large numbers that when they suddenly die off en masse, the breaking down of their cells by bacteria destroys the amount of dissolved oxygen in the water, hurting the animals and plants that live there.

Fungi

Fungi are organisms that scientists once confused with plants. However, scientists have found that, at the cell level, the fungi are more like animals than they are like plants. For one thing, fungi cannot synthesize their own food like plants do, but instead they eat other organisms as do animals. Fungi come in a variety of shapes and sizes and different types. They can range from single cells to enormous chains of cells that can stretch for miles. Fungi include single-celled living things that exist individually, such as yeast and multicellular clusters, such as molds or mushrooms. Yeast cells look round or oval under a microscope. They're too small to see as individuals, but you can see large clusters of them as a white powdery coating on fruits and leaves.

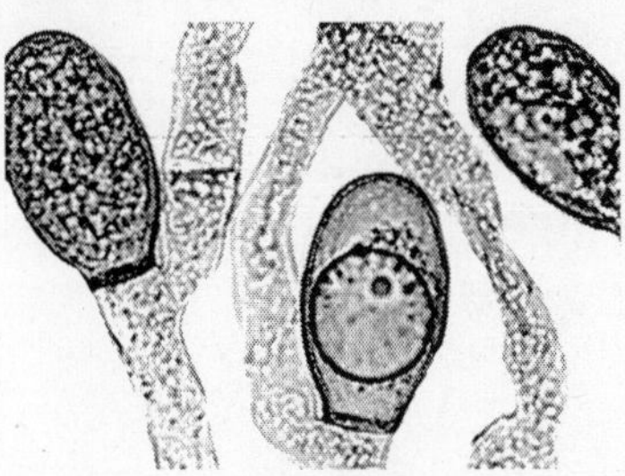

Fig. Fungi

Molds are described as filament-like because they form long filament-like, or thread-like, strands of cells called hyphae. These hyphae are what give mold colonies their fuzzy appearance. They also form the fleshy body, or mushroom, that some species grow. It may seem strange to think of something as big as a mushroom as a microbe. But the cells of the hyphae making up that mushroom are connected in a closer way than the cells of other multicellular living things, like you and me, are. Fungi usually grow best in environments that are slightly acidic. They can grow on substances with very low moisture. Fungi live in the soil and on your body, in your house and on plants and animals, in freshwater and seawater. A single teaspoon of topsoil contains about 120,000 fungi.

Fungi are basically stationary. But they can spread either by forming reproductive spores that are carried on wind and rain or by growing and extending their hyphae. Hyphae grow as new cells form at the tips, creating even longer chains of cells. Fungi absorb nutrients from living or dead organic matter that they grow on. They absorb simple, easily dissolved nutrients, such as sugars, through their cell walls. They give off special digestive enzymes to break down complex nutrients into simpler forms that they can absorb.

Some fungi are quite useful to us. We've used several kinds to make antibiotics to fight bacterial infections. These antibiotics are based on natural compounds the fungi produce to compete against bacteria for nutrients and space. We use baker's yeast, to make bread rise and to brew beer. Fungi break down dead plants and animals and keep the world tidier. We're exploring ways to use natural fungal enemies of insect pests to get rid of these bugs. There are some dangerous fungi that cause diseases in plants, animals and people. Fungi ruin about a quarter to half of harvested fruits and vegetables each year.

MICROBES AND CHEMICAL CHANGES

At about this time Liebig the German chemist, had claimed that fermentation and putrefaction were processes caused by the chemical instability of organic molecules. On the other hand, Pasteur had shown that living yeast cells were required for the conversion of sugar into alcohol and had suggested that these fermentations and transformations are brought about by the agency of microorganisms. At this time. the distilleries in France, where alcohol was produced from sugar beet had difficulties in their fermentations and requested Pasteur for help. Pasteur found that the trouble was due to the replacement of the alcoholic fermentation by a lactic fermentation. Instead of finding yeast in the fermenting brew, he found rod-shaped structures.

He concluded that these rod shaped organisms were responsible for the conversion of sugar into lactic acid. Subsequently, he also showed that different fermentations are brought about by specific microorganisms. For example, the alcoholic fermentation was the result of living yeast cells while the lactic acid fermentation was the result of the activity of lactic bacteria. Interestingly in 1897, after Pasteur, Buchner showed that intact yeast cells were not. required for the conversion of sugar into alcohol and carbon dioxide He prepared a cell free yeast juice by grinding yeast cells with sand and showed that this cell free juice had the ability to carry out the alcoholic fermentation. He concluded that the active agent in the yeast cells responsible for bringing about the fermentation was a protein and he named it as zymase. This finding gave support to Liebig's thesis that living cells were not required to bring about fermentations and marks the beginning of the study of enzyme chemistry.

ANCIENT MICROBIOLOGICAL HISTORY

Ancient man recognized many of the factors involved in disease. Early civilizations on crete, india, pakistan and scotland invented toilets and sewers; lavatories, dating around 2800 bc, have been found on the orkney islands and in homes in pakistan about the same time. One archaeologist has stated that "the high quality of the sanitary arrangements could well be envied in many parts of the world today". In rome, 315 ad, the public lavatories were places where people routinely socialized and conducted business.

Ten to twenty people could be seated around a room, with their wastes being washed away by flowing water; it must have been difficult to *"stand on your dignity"* under such circumstances. The chinese used toilet paper as early as ad 589. In europe moss, hay and straw were used for the same purpose. I can personally attest to the use as late as 1962 of "slick magazines" as toilet paper in certain european camp grounds. The first cities to use water pipes (of clay) were in the indus valley of pakistan around 2700 bc. Metal water pipes were used in egypt (2450 bc) and the palace of knossos on crete around 2000 bc had clay pipes.

Rome built elaborate aqueducts and public fountains throughout its empire to insure a clean supply of water for its citizens. Rome had a "water commissioner" who was responsible for seeing that the water supply was kept adequate and clean; the punishment for pollution of the water supply was death. Lead was commonly used for roman pipes and the subsequent fall of the roman empire has been related by some to the effects of lead on the roman brain.

Most ancient peoples recognized that some diseases were communicable and isolated individuals thought to carry "infections". An example of this is the universal shunning of lepers, which occurs even today. When the black death struck europe, entire villages were abandoned as people fled in an effort to escape the highly infectious plague. Similarly, in the middle ages the rich of europe fled to their country homes when small pox struck in an effort to escape its terrible consequences. The fact that people who recovered from a particular disease were immune to that disease was probably recognized many different times in many places. Often these survivors were expected to nurse the ill. Greek and roman physicians routinely prescribed diet and exercise as a treatment for ills. Sadly, we know that this knowledge did not help most of our ancestors and that the human life span was, until the last 200 years, more often than not cut short due to infectious disease. Even today approximately 15,000,000 children die per year, mainly from infectious diseases that are preventable with basic sanitation, immunization and simple medical treatments. One might honestly question just how far we have come in our treatment of disease. An excellent synopsis of the history of microbiology ancient people had certainly seen masses of microbes, such as mold and bacterial colonies, on spoiled food, but it is doubtful if anyone considered that they were viewing living organisms.

Small boys and maybe a few love-sick adults staring into a clear pond, must have seen tiny specks moving rapidly about and some may have considered them living creatures, but to express this to their friends would be equivalent to us telling our friends that we'd seen a flying saucer. The first person to report seeing microbes under the microscope was an englishman, robert hooke. Working with a crude compound microscope he saw the cellular structure of plants around 1665. He also saw fungi which he drew. However, because his lens were of poor quality he was apparently unable to "see" bacteria.

Anton van leeuwenhoek was a man born before his time. Although not the first to discover the microscope or to use magnifying lens, he was the first to see and describe bacteria. We know that he was a "cloth merchant" living in delft holland. And that he used magnifying lens to view the quality of the weave of the merchandise he purchased. He traveled to england in 1668 to view english cloth and there he saw drawings of magnifications of cloth much greater than any of the current lens available in holland would do.

He returned to holland and took up lens grinding. Being meticulous, he developed his lens grinding to an art and in the process tested them by seeing how much detail he could observe with a given lens. One can guess that he chanced to look at a sample of pond water or other source rich in microbes and was amazed to see distinct, uniquely shaped organisms going, apparently purposefully, about their lives in a tiny microcosm. He made numerous microscopes from silver and gold and viewed everything he could including the scum on his teeth and his semen.

His best lens could magnify ~300-500 fold which allowed him to see microscopic algae and protozoa and larger bacteria. He clearly had excellent eyesight because he accurately drew pictures of microbes that were at the limit of the magnification of his lens. He used only single lens and not the compound lens of the true microscopes we employ today; which makes his observations all the more amazing. He wrote of his observations to the royal society of london in 1676 and included numerous drawings. He astonished everyone by claiming that many of the tiny things he saw with his lens were alive because he saw them swimming purposefully about. This caused no end of shock and wonderment and numerous people hurried to delft to see if this dutchman was "in his cups" or if he was really onto something new and wonderful. A few minutes with one of his numerous microscopes was all it took to convert his visitors to enthusiastic believers in the existence of these tiny beasties living all around them. His discovery was the equivalent of our finding life on mars today.

THE CLEAN NUT AND THE REVENGE OF THE BACTERIA

In the 1800s people began to use hospitals. Hospitals also became centers of physician training. In 1841 young doctor ignaz semmelweis was hired to run a maternity ward in a vienna hospital. There were two birthing wards in his preview, one run by midwives and the other by doctors. Semmelweiss noticed that the death rate among mothers in the doctor's ward ran as high as 18 per cent from the blood infection known as child bed fever or puerperal sepsis, whereas in the midwife ward the death rate was much lower. When he suggested that the doctors might contribute to this he was fired. Subsequently rehired, he saw a friend die from puerperal sepsis after cutting himself during an autopsy of a patient who had died of puerperal sepsis. He reasoned that there was an invisible agent that caused both deaths and that one could transfer it from the autopsy room to the birthing rooms and thus infect the mothers during birthing.

Acting on this assumption, semmelweis instituted sanitary measures which included having the doctors wash their hands in disinfectant and change from lab coats dripping with pus and blood from the autopsy room to clean lab coats before examining patients or assisting in a birth. The death rate of the mothers dropped by 2/3 in his ward. However, the other doctors objected so strongly to his rules that semmelweis was again fired and left vienna. He took other hospital jobs where he instituted the same standards of cleanliness which resulted in the same decline in deaths and a revolt of his fellow physicians. Semmelweis ended up dying in an insane asylum from a blood infect that resembled puerperal sepsis.

SPONTANEOUS GENERATION

The mystery of life has puzzled and confounded humans since the first human began to contemplate his world. The religions of ancient societies were built around the seasons, the sun and the renewal of life as these were so clearly tied to survival; both of the human species through birth and death and of the individual in the attainment of sustenance. Spontaneous generation or the idea that life routinely arises from non-life was a common sense explanation of the miracle of life.

It had the advantage of simplicity, ease of under-standing and didn't require any waste-of-time thinking. As science and the scientific method grew with the slow accumulation of knowledge, observant individuals began to consider the origin of life more deeply. Simple observations convinced many people that all the larger animals and plants produced life from previous life. Despite this, the mass of humans clung to the comfortable idea of spontaneous generation further, religions saw it as a convenient way to demonstrate the hand of god operating continuously in the world. Some individuals, such as j.b. Van helment even described how one could make mice from grain, a jar and dirty rags by putting them together in a dark place for a few weeks and soon mice would appear in the jar.

Other, more perceptive individuals, like f. Redi tested the common idea that maggots arise via spontaneous generation on rotting meat. He placed a piece of meat in three jars, one he left open, one he corked tightly and the third he covered with a fine mesh gauze. Maggots only appeared in the open container, no matter how long he left the jars. Redi's experiment was important because of its eloquent simplicity. Anyone could repeat it and obtain the same clear results. Nevertheless, many people clung fiercely to the idea of spontaneous generation, while others designed experiments to test it. In every case the results of the majority of these experiments indicated that spontaneous generation did not occur. The intellectual ferment this controversy stirred up gradually evolved into the scientific method as the various antagonists questioned each other's assumptions and, more importantly, their experimental design.

These arguments forced the designing of better experiments and eventually persuaded all but the most recalcitrant believers to discard spontaneous generation as an explanation for all higher life forms. Then, in the 1800's the refinement of the microscope, through which people could see tiny life forms that they assumed were simple, gave spontaneous generation proponents new life.

Again, flawed common sense led reasonable people astray. As the existence of microscopic life was accepted, the assumption was that such life must be simple compared to higher, more complex life. The reasoning that followed this erroneous assumption was that since the microbes were small they must be simple & it followed that they were formed by spontaneous generation, hence god was still at work creating micro-life. Small is not simple! The battle over spontaneous generation raged anew both from the pulpit and the lab. A number of scientists performed elementary experiments in which they treated soups and broth's, which left unheated would team with microbes after a few days, with heat to destroy any life present in them and asked the question: "Would new life arise in these sterile soups"? Spallanzani boiled "Soup" in glass containers and melted the glass closed. The observation that nothing subsequently grew in this "heated" soup suggested that spontaneous generation didn't work. His detractors, rightly criticized his experiments, proposing that since air is necessary for life and since he had sealed the flask to air, obviously no life could develop. Others boiled soups and microbes grew, thus apparently supporting spontaneous generation. But again the preponderance of data suggested that spontaneous generation did not even apply to the "simple" microbes. In 1859 one of the fathers of modern microbiology, l. Pasteur decided to settle the question of spontaneous generation once and for all. A genius at devising definitive experiments, pasteur first drew the necks of glass flasks out so that they remained open to the air, but were bent so that air could only enter by a curved path.

He then added broth and boiled it to destroy contaminating microbes. These flasks were then incubated and observed for months. He reasoned that the microbes in the air that could contaminate the sterile broth would be trapped on the sides of the thin glass necks before they reached the sterile broth. If spontaneous generation didn't occur no growth should take place. This is exactly what happened, the flasks remained sterile indefinitely, until pasteur tipped the sterile broth up into the curved neck where he predicted the airborne organisms would have settled. After doing this the broths always grew microbes. These experiments ended the spontaneous generation controversy because these experiment was so elegant and simple and the results so clear, that anyone could repeat them.

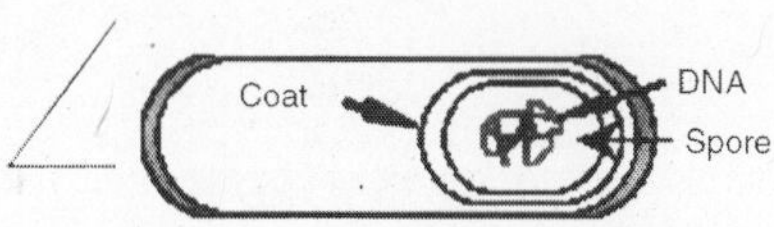

Fig. Spore Structure

Later, an earlier problem, in which occasional heated-broths did not remain sterile, was explained with the discovery of the heat resistant bacterial some of which could can survive several hours of boiling without being killed.

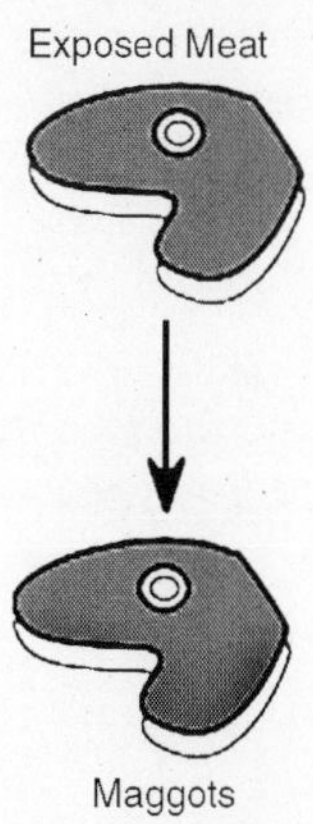

Fig. The Redi Experiment

Pasteur discovered many of the basic principles of microbiology and, along with R. Koch, laid the foundation for the science of microbiology. In 1857 Napoleon III was having trouble with his sailors mutinying because their wine was spoiling after only a few weeks at sea. Naturally Napoleon was distraught because his hopes for world conquest were being scuttled (pardon the pun) over a little spoiled wine, so he begged Pasteur for help.

Pasteur, armed with his trusty microscope, accepted the challenge and soon recognized that by looking at the spoiled wines he could distinguish between the contaminants that caused the spoilage and even predict the taste of the wine solely from his microscopic observations. He then reasoned that if one were to heat the wine to a point where its flavour was unaffected, but the harmful microbes were killed it wouldn't spoil. As we are aware this process, today known as pasteurization, worked exactly the way he predicted and is the foundation of the modern treatment of bottled liquids to prevent their. It is important to realise that pasteurization is NOT the same as sterilization. Pasteurization only kills organisms that may spoil the product, but it allows many microbes to survive, whereas sterilization kills all the living organisms in the treated material.

Pasteur also realized that the yeast that was present in all the produced the alcohol in wine. When he announced this, a number of famous scientists were enraged, because the current theory of wine production was that wine formation was the result of SPONTANEOUS chemical changes that occurred in the grape juice. Pasteur was attacked furiously at scientific meetings, to the point where certain scientists did humorous skits about Pasteur and his tiny little yeast "stills" turning out alcohol. Pasteur had the last laugh however as people all over the world soon realized that if he was right they could control the quality of wine by controlling the yeast that made it. In a short period many others verified his observations and the opposition sank without a sound.

Pasteur discovered many of the basic principles of microbiology and, along with R. Koch, laid the foundation for the science of microbiology. In 1857 Napoleon III was having trouble with his sailors mutinying because their wine was spoiling after only a few weeks at sea. Naturally Napoleon was distraught because his hopes for world conquest were being scuttled (pardon the pun) over a little spoiled wine, so he begged Pasteur for help. Pasteur, armed with his trusty microscope, accepted the challenge and soon recognized that by looking at the spoiled wines he could distinguish between the contaminants that caused the spoilage and even predict the taste of the wine solely from his microscopic observations. He then reasoned that if one were to heat the wine to a point where its flavour was unaffected, but the harmful microbes were killed it wouldn't spoil. As we are aware this process, today known as pasteurization, worked exactly the way he predicted and is the foundation of the modern treatment of bottled liquids to prevent their. It is important to realise that pasteurization is NOT the same as sterilization. Pasteurization only kills organisms that may spoil the product, but it allows many microbes to survive, whereas sterilization kills all the living organisms in the treated material.

Pasteur also realized that the yeast that was present in all the produced the alcohol in wine. When he announced this, a number of famous scientists were enraged, because the current theory of wine production was that wine formation was the result of SPONTANEOUS chemical changes that occurred in the grape juice. Pasteur was attacked furiously at scientific meetings, to the point where certain scientists did humorous skits about Pasteur and his tiny little yeast "stills" turning out alcohol. Pasteur had the last laugh however as people all over the world soon realized that if he was right they could control the quality of wine by controlling the yeast that made it. In a short period many others verified his observations and the opposition sank without a sound.

DEVELOPMENT OF MICROBIAL KINETICS

Microbial growth can be defined as an orderly increase in cellular components, resulting in cell enlargement and eventually leading to cell division. This definition is not strictly accurate as it implies that a consequence of growth is always an increase in cell numbers. However, under certain conditions growth can occur without cell division, for example, when cells are synthesizing storage compounds, *e.g.* glycogen or poly-b-hydroxybutyrate. In this situation the cell numbers remain constant, but the concentration of biomass continues to increase. This is also true for coenocytic organisms, such as some fungi, that are not divided into separate cells. Their growth results only in increased Size.

BACTERIAL GROWTH KINETICS

Growth kinetics of homogeneous unicellular suspension. cultures can be modeled using differential equations in a continuum model. However, filamentous growth and growth in heterogeneous cell aggregates and assemblage, particularly biofilms, colonies, flocs, mats and pellicles, is much more complex. In fact, heterogeneous systems require a very different approach using cellular automaton and Swarm system models, *e.g.* BacSim. The growth kinetics of filamentous organisms and heterogeneous systems will not be discussed here.

MODELING THE MICROBIAL GROWTH KINETICS

The growth model that will be examined is bacterial binary fission in homogeneous suspension cultures, where cell division produces identical daughter cells. Each time a cell divides is called a generation and the time taken for the cell to divide is referred to as the generation time. Therefore, the generation time or doubling time *(td)* is the time required for a microbial population to double. Theoretically, after one generation, both the microbial cell population and biomass concentration have doubled. However, as previously stated, under certain conditions growth can be associated with an increase in biomass and not cell numbers. Also, the generation time recorded during microbial growth is in reality an average value, as the cells will not be dividing at exactly the same rate. At anyone time there are cells at different stages of their cell cycle. This is termed asynchronous growth. However, under certain conditions synchronous growth can be induced so that all cells divide simultaneously, which is a useful research tool in the study of microbial physiology. Microbial fermentations in liquid media can be carried out under different operating conditions, *i.e.* batch growth, fed-batch growth or continuous growth. Batch growth involves a closed system where all nutrients are present at the start of the fermentation within a fixed volume. The only further additions may be acids or bases for pH control, or gases (*e.g.* aeration, if required). In fed-batch systems fresh medium or medium components are fed continuously, intermittently or are added as a single supplement and the volume of the batch increases with time. Continuous fermentations are open systems where fresh medium is continuously fed into the fermentation vessel, but the volume remains constant as spent medium and cells are removed at the same rate.

BATCH GROWTH

During batch fermentations the population of microorganisms goes through several distinct growth phases: lag, acceleration, exponential growth, deceleration, stationary and death. In the lag phase virtually no growth occurs and the microbial population remains relatively constant. Nevertheless, it is a peripd of intense metabolic activity as the microbial inoculum adapts to the new environment. When cells are inoculated into fresh medium they may be deficient in essential enzymes, vitamins or cofactors, etc., that must be synthesized in order to utilize available nutrients, prior to cell division taking place.

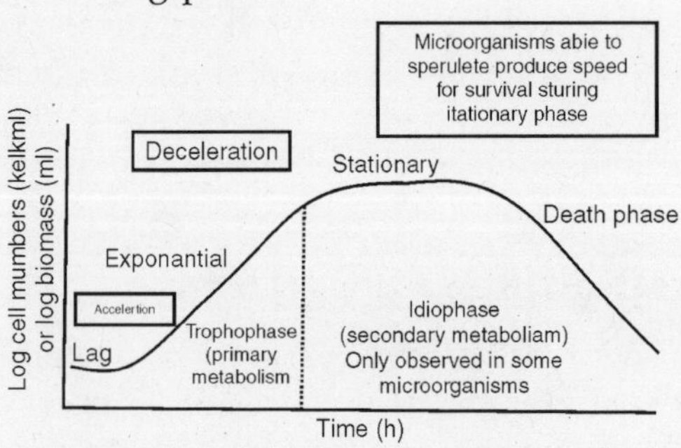

The chemical composition of the fermentation media influences the length of the lag phase. It is usually longer if the inoculum was grown up using a carbon source different from that in the fresh medium, because the cells must synthesize enzymes required to catabolize the new substrate. Physiological stress may also have an effect, especially as cells are often transferred from an inoculum medium of low osmotic pressure (low solute concentration) to fresh medium of higher osmotic pressure (high solute concentration). Other factors influencing the length of the lag phase are the age, concentration, viability and morphology of the inoculum.

Generally, inocula prepared from cells harvested in the exponential growth phase (period of most rapid growth) exhibit shorter lag phases than those harvested from subsequent stages. Once the cells have adapted to their new environment they enter the acceleration phase. Cell division occurs with increasing frequency until the maximum growth rate (mmax) for the specific conditions of the batch fermentation is reached.

At this point exponential growth begins and cell numbers/biomass increase at a constant rate. Mathematically, this exponential growth can be described by two methods; one is related to biomass *(x)* and the other to cell numbers *(N)*. For cell biomass, growth can be considered as an autocatalytic reaction. Therefore, the rate of growth is dependent on the biomass concentration, *i.e.* catalyst, that is present at any given time. This can be described as follows:

Rate of change of biomass is $dx/dt = \mu x$ (5.1)

where x = concentration of biomass (g/L),

μ= specific growth rate (per hour) and t= time (h).

When a graph is plotted of cell biomass against time, the product is a curve with a constantly increasing slope. Equation can also be rearranged to estimate the specific growth rate (μ): m= $1/x$ * $dxldt$ During any period of true exponential growth, equation can be integrated to provide the following equation: xt=xoemt where xt = biomass concentration after time' t, xo = biomass concentration at the start exponential growth, and e = base of the natural logarithm.

Taking natural logarithms, loge (In), gives

$$\text{In } Xt = \text{In Xo} + \mu t$$

This equation is of the form y = c (intercept on y axis) + mx where m = gradient, which is the general equation for a straight-line graph. For cells in exponential phase, a plot of natural log of biomass concentration against time, a semilog plot, should yield a straight line with the slope (gradient) equal to μ, or m= (In xt-In xo)/t (note: when plotting log10 values instead of the natural log, the gradient of the semilog plot is equal to m/*2.303;*).

A second approach is to examine growth in relation to cell number, where the number of cells at the start of exponential growth is No. If we consider a hypothetical case where the microbial cell population at this point of exponential growth is one (No=1), we can describe binary fission as follows:

No. of divisions 0 1 2 3 n

No. of cells 1 2 4 8 $2n$

Mathematically No N 2 No 2 * 2 No 2 * 2 * 2-

No21 *No22 No23* No $2n$

Consequently, after a period of exponential growth, time *(t)*, the number of cells *(Nt)* is given by

Nt = No $2n$

where n=the number of divisions, No = initial cell number.

Taking natural logarithms gives

$$\text{In } Nt = \text{lnNo} + \text{nln2}$$

Therefore, the number of divisions *(n)* that have taken place is given by In Nt-In No n = In2

The number of divisions per unit time during this period of exponential growth is determined by dividing by the time period *(t):*

$$n = \text{lnNt-lnNo } t \ t\text{ln2}$$

where n/t = division rate constant (average number of generations per hour).

Often, we are not really interested in the number of divisions per hour, unit time, but rather in the mean generation time or doubling time *(td)*, that is, the time required to undergo a single generation that doubles the population. Thus,

$$td = t = t \text{ ln2 } n \text{ In } Nt\text{-In No} \quad (5.10)$$

During exponential growth, when all nutrients are supplied in excess and are therefore non-limiting, there is a direct relationship between cell numbers and biomass concentration, assuming that mean cell size is constant. This is balanced growth, and a direct relationship between specific growth rate and doubling time can also be established. However, under conditions where an essential nutrient becomes limiting, unbalanced growth arises, and variations in cell numbers *(N)* and biomass *(x)* concentration occur, as during the synthesis of cell storage compounds.

If we consider a situation where at time zero, the cell biomass is *xo,* then after a fixed period of time *(t)* of exponential growth, equivalent to one doubling time *(td)*, the microbial biomass will double to *2xo, i.e. xt=2xo,* when *t=td,* Substituting these parameters into equation gives *2xo* = *xoe* mt d

Taking natural logarithms produces

$$\text{In}2xo = \text{In } xo + \text{mtd}$$

or

$$\mu\text{td} = \text{In } 2.$$

Therefore, in this case

$$\text{td} = 0.693\mu$$

Equation, rate of change of biomass *(dx/dt*=mx*)*, predicts that growth will occur indefinitely. However, during batch growth the microorganisms are continuously metabolizing the finite supply of nutrients available in the fermentation broth. After a certain time the growth rate decreases and eventually stops. This cessation of growth can be due to depletion of essential nutrients (carbon source, essential amino acids, etc.) or the build-up of toxic metabolites, such as ethanol and lactic acid, or a combination of nutrient depletion and toxin accumulation.

Monod showed that growth rate is an approximate hyperbolic function of the concentration of the growth-limiting nutrient(s). This impact of essential nutrient depletion on-growth can be described mathematically by the Monod equation, in a form similar to that used in biochemistry, where Michaelis-Menten kinetics define the rate of an enzymecatalysed reaction in relation to its substrate concentration.

$$\mu = \frac{\mu_{max} S}{Ks + S}$$

where μ_{max} = maximum specific growth (per hour) of the cells, *i.e.* when substrate concentration is not limiting; S = concentration of limiting substrate (g/L); *Ks* = saturation constant, concentration (g/L) of limiting nutrient enabling growth at half the maximum specific growth rate, *i.e. m.*=1/2 mmax and is a measure of the affinity of the cells for this nutrient.

When a microorganism is provided with the limiting substrate at a concentration much greater than the *Ks,* and with all other nutrients in excess, the microorganism will grow exponentially at its maximum rate, *i.e.* when S» *Ks,* then $\mu.=\mu._{max}$. However, as the level of this substrate decreases, it eventually becomes limiting and can no longer sustain μ_{max}.

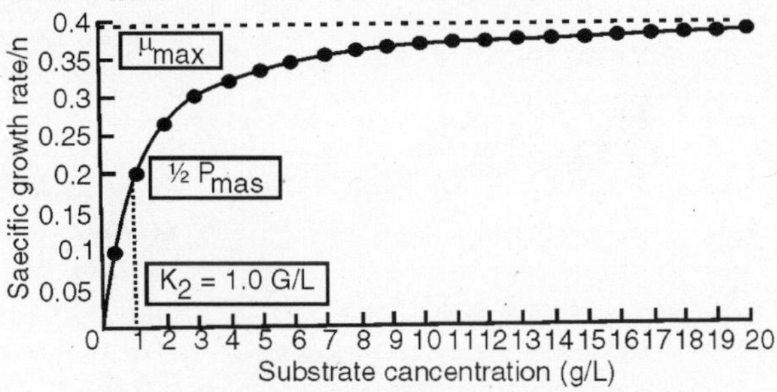

Fig. The Relationship Between Substrate Concentration and specific growth rate.

This is the beginning of the deceleration phase. As the residual concentration of the limiting substrate approaches *Ks* and then falls below this concentration, there is an accompanying gradual decrease in growth rate (μ). The growth rate of a microorganism with a very high affinity for a rate-limiting substrate (*i.e.* a low *Ks)* will not be affected until the substrate concentration becomes very low. However, where there is a low affinity for the limiting substrate (*i.e.* a high *Ks),* the growth rate will begin to fall even at relatively high substrate concentrations and the organism exhibits a longer deceleration phase. The specific growth rate of the microorganism continues decelerating until all of the available limiting substrate is metabolized. Growth is no longer sustainable and the cells enter the stationary phase. At this point, the overall growth rate has declined to zero and there is no net change in cell numbers/biomass (rate of cell division equals rate of cell death). However, the microorganisms are still metabolically active, involved in metabolizing intracellular storage compounds, utilizing nutrients released from lysed cells, and in some cases producing secondary metabolites.

The duration of the stationary phase varies depending on the microorganisms involved and the environmental conditions. For cells unable to survive by forming spores, this is followed by an exponential death phase when the cells die at a constant rate and often undergo lysis.

APPLICATION OF BATCH FERMENTATIONS

During batch fermentations certain environmental conditions continually change, particularly nutrient and product concentrations, as does the specific growth rate, because the cells must pass through the sequence of growth phases described above.

Consequently, the system never achieves steady-state conditions. A further disadvantage is that several distinct practical stages are associated with the operation of a batch fermentation:

- Charging of the fermenter with fresh medium;
- Sterilization of the fermenter and medium;
- Inoculation of the fermenter;
- Production of microbial products;
- Harvesting of biomass and spent fermentation broth; and finally
- Cleaning of the vessel.

This has major economic implications for industrial processes. For a considerable period of time, the fermentation vessel is not producing microbial products, but is being cleaned, filled, sterilized, etc. The non-productive period is referred to as the down-time of the fermenter.

OPTIMIZATION OF BATCH FERMENTATIONS

During the development of a batch process, key growth parameters can be determined that enable the production of a given microbial product to be optimized, whether it is the biomass itself or a specific metabolite. An important parameter is the yield coefficient (Y), which is determined on the basis of the quantity of rare limiting nutrient, normally the carbohydrate source, converted into the microbial product. In respect of biomass, it is defined as,

$$x = Y\, x/s(S\text{-Sr})$$

where x = biomass concentration (g/L), $Y\, x/s$ =*yield* coefficient (g biomass/g substrate utilized), S = initial substrate concentration (g/L), and Sr = residual substrate concentration (g/L). In the case of biomass production, the yield coefficient relates to the quantity of biomass produced per gram of substrate utilized. Therefore, the higher the yield coefficient, the greater the percentage of the original substrate converted into microbial biomass.

For microbial metabolic products *(p)* the yield coefficient is related to the quantity of metabolite produced in relation to the quantity of substrate used *Y p/s)'* Determination of yield coefficients is vitally important because the cost of the fermentation medium, particularly the carbon source, can be a significant proportion of the overall production cost. By performing a range of experiments under different operating conditions, varying medium constituents and component concentrations, pH, temperature, etc., optimum growth/ production conditions can be established.

It is also important to determine the maximum specific growth rate (mmax) of the production organism. This is particularly true for primary metabolites, where product formation is related to growth. To optimize the overall productivity of the system, the microorganism must usually be grown at its mmax. As previously stated, the operating substrate concentration has a major effect on the growth rate of a microorganism. By performing a series of batch fermentations, each with a different initial concentration of the limiting substrate, the specific growth rate (m) for each experiment can be determined. These data can then be used to estimate both mmax and the saturation constant *(Ks)* by simply taking the reciprocal values in the Monod equation and rearranging equation to give,

$$1 = Ks + S$$

$$\mu \ \mu_{max}S$$

or

$$1 = Ks + S$$

$$\mu \ \mu_{max}S \ \mu_{max}S$$

and

$$1 = Ks * 1 + S$$

$$\mu \ \mu_{max}S \ \mu_{max}S$$

A plot of 1/m against 1/*S* should produce a straight line with the intercept on the y-axis at $1/\mu_{max}$ and a gradient equal to $K/\ \mu_{max}$. Therefore, an the key kinetic parameters can be readily determined and when the values of *K*, mmax and Y are known, a complete quantitative description can be given of the growth events occurring during a batch culture.

CONTINUOUS GROWTH KINETICS

Continuous culture fermentations have many applications for both laboratory research and industrial-scale processes. Studies can be performed on all aspects of cell growth, physiology and biochemistry. They are useful for ecological studies and as a genetics tool for the examination of mutation rates, mutagenic effects, etc. Application in industrial fermentations overcomes many limitations of batch processes. Initially, continuous fermentations start as batch cultures, but exponential growth can then be extended indefinitely, in theory, through the continuous addition of fresh fermentation medium. The reactor is continuously stirred and a constant volume is maintained by incorporating an overflow weir or other levelling device.

Fresh medium is continuously added and displaces an equal volume of spent fermentation broth and cells at the same rate as fresh medium is introduced. Steadystate conditions prevail, where the rate of microbial cell growth equals the rate at which the cells are displaced from the vessel. As with batch fermentations, the specific rate at, which the microorganism grows in continuous culture is controlled by the availability of the rate-limiting nutrient.

Therefore, the rate of addition of fresh medium controls the rate at which the microorganisms grow. However, the actual rate of growth depends not only on the volumetric flow rate of the medium into the reactor, but also on the dilution rate *(D).* This equals the number of reactor volumes passing through the reactor per unit time and is expressed in units of reciprocal time, per hour.

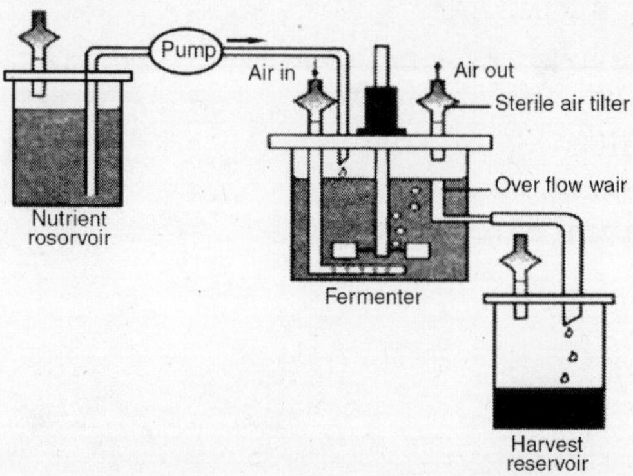

Fig. Continuous Culture Apparatus.

$$D = F\ V$$

where D = dilution rate (per hour), *F* = flow (L/h) and V= reactor volume (L). The term D is the reciprocal of the mean residence time or hydraulic retention time, as used in waste-water treatment.

Addition of fresh medium into the reactor can be controlled at a fixed value, therefore the rate of addition of the rate-limiting nutrient is constant. Within certain limits, the growth rate and the rate of loss of cells from the fermenter will be determined by the rate of medium input. Therefore, under steady state conditions the net biomass balance can be described as,

dx = rate of growth-rate of loss from reactor

dt in reactor (wash-out) or

$$dx = \mu x - Dx$$

Under steady-state conditions the rate of growth = rate of loss, hence *dx*/*dt*= 0 and therefore,

$$\mu x = Dx \textit{ and } m \doteq D$$

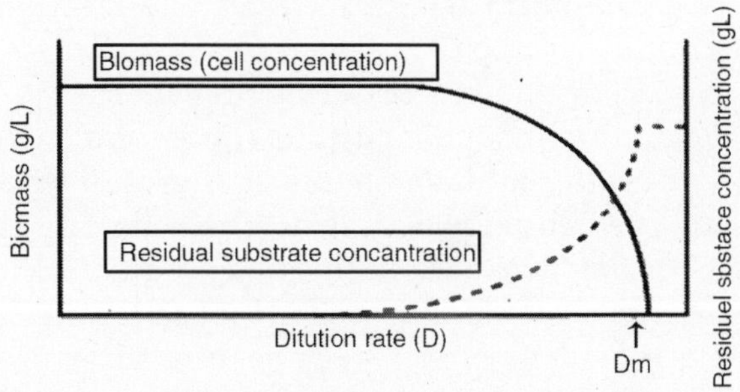

Fig. Growth of a Microorganism in Continuous Chemostat culture.

Consequently, at fixed flow rates and dilution rates under constant physical and chemical operating conditions, *i.e.* under steady-state conditions, the specific growth rate of the microorganism is dependent on the operating dilution rate, up to a maximum value equal to μ_{max}.

If the dilution rate is increased above μ_{max} complete wash-out of the cells occurs, as the cells have insufficient time to 'double' before being washed out of the reactor via the overflow. The point at which this is just avoided is referred to as the critical dilution rate (Dcri). For any given dilution rate, under steady-state conditions, the residual substrate concentration in the reactor can be predicted by substituting D for *ll* in the Monod equation:

$$D = \frac{\mu_{max}\, Sr}{Ks + Sr}$$

where Sr is the steady-state residual substrate concentration in the reactor at a fixed dilution rate. Rearrangment gives,

$$D(Ks + Sr) = \mu_{max}\, Sr \text{ or dividing by Sr then gives}$$

$$D\,Ks + D\,Sr = \mu_{max}\, Sr$$

$$D\,Ks + D = \mu_{max}\, Sr$$

Hence,

$$Sr = \frac{DKs}{\mu_{max} - D}$$

Consequently, the residual substrate concentration in the reactor is controlled by the dilution rate. Any alteration to this dilution rate results in a change in the growth rate of the cells that will be dependent on substrate availability at the new dilution rate.

Thus, growth is controlled by the availability of a rate-limiting nutrient. This system, where the concentration of the ratelimiting nutrient entering the system is fixed, is often described as a chemostat, as opposed to operation as a turbidostat, where nutrients in the medium are not limiting. In this case turbidity or absorbance of the culture is monitored and maintained at a constant value by regulating the dilution rate, *i.e.* cell concentration is held constant.

At low dilution rates with fixed substrate concentrations, the residual substrate concentration will be low. However, as D approaches μ_{max} the residual substrate concentration increases along with the growth rate of the microorganism. Beyond Dcrit, input substrate concentration will equal output concentration, as all the cells have been lost from the system. Thus, this continuous reactor can be described as a selfregulating nutrient-limited chemostat.

The concentration of biomass or microbial metabolite in a continuous fermenter under steady-state conditions can be related to the yield coefficient, as described in the batch fermentation section. Inserting the equation for residual substrate into the biomass or metabolic product yield coefficient equation gives, in this case for steadystate biomass *(x)*,

$$x = Yx/s\ \mathrm{SR} - DKs$$

$$\mu_{max} - \mathrm{D}$$

where SR is the substrate concentration of inflowing medium or

$$x = Yx/S(SR\text{-}\mathrm{Sr})$$

Therefore, the biomass concentration under steady state conditions is controlled by the substrate feed concentration and the operating dilution rate. Under non-inhibitory conditions, where there is no substrate or product inhibition, the higher the feed concentration, the greater the biomass concentration and residual substrate concentration remains constant. However, the higher the dilution rate, the faster the cells grow, which results in a simultaneous increase in the residual substrate concentration and a consequent reduction in the steady-state biomass concentration. As D approaches μ_{max}, the biomass concentration becomes even lower, yet the cells grow faster and there is a concurrent increase in the residual substrate concentration.

MICROBIAL BIOMASS AND NITROGEN CYCLE

THE AQUATIC NITROGEN CYCLE

The aquatic nitrogen cycle is a complex system consisting of several different forms of nitrogen with different oxidation states that are interconnected by numerous conversion reactions. The different components of the aquatic nitrogen cycle can be separated from each other in space and/ or time. The nitrogen cycle plays an important role in aquatic systems since nitrogen is usually the element limiting microbial production in these systems. However, due to its complexity, various aspects of the nitrogen cycle are still poorly understood.

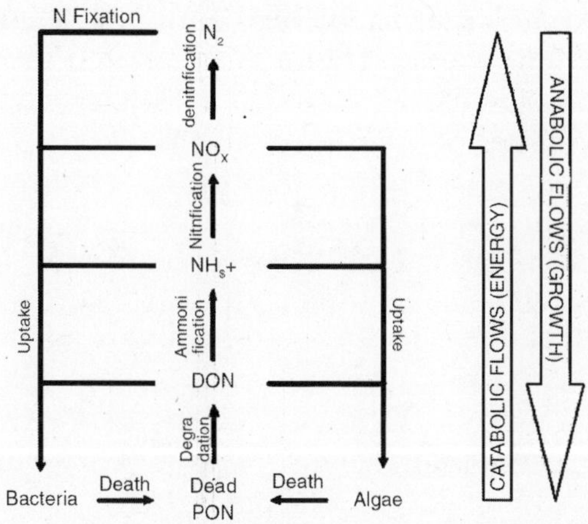

Fig. General Overview of Nitrogen Transformation Processes within the Nitrogen Cycle, Including Nitrogen flows Through Bacteria and algae.

Bioavailable nitrogen is present in the environment as "free" nitrogen (atmospheric N_2) and "fixed" nitrogen. The latter can roughly be divided into particulate nitrogen (functionally defined as all nitrogenous material that can be captured by filtration) and dissolved nitrogen (all nitrogenous material that passes through a filter, typically GF/F).

The dissolved fixed nitrogen pool can be subdivided into dissolved inorganic nitrogen (DIN) and dissolved organic nitrogen (DON). The DIN pool includes ammonium (NH^{4+}), nitrate (NO^{3-}) and nitrite (NO^{2-}) of which the last two can be taken together as NOx-. The DON pool is a more complex pool of which only a small fraction (< 20 per cent) has been chemically identified (Benner 2002, Bronk 2002). The identifiable fraction mainly consists of urea, dissolved free amino acids (DFAAs) and dissolved combined amino acids (DCAAs, mainly proteins and peptides). The conversion reactions within the nitrogen cycle can have two "Directions". In catabolic (or "Dissimila-tory") reactions, nitrogenous substrates are used as a source of energy, these reactions are generally mediated by bacteria. In anabolic (or "assimilatory") reactions, nitrogenous substrates are used as a source of nitrogen for growth (*i.e.* production of biomass), which requires energy. The uptake and incorporation of various forms of fixed nitrogen by natural aquatic microbial communities.

Microbial Nitrogen Uptake

Natural aquatic microbial communities (mainly bacteria and algae) can use various forms of nitrogen as nitrogen sources for growth. Although N_2 is abundantly present in the atmosphere, its importance as a nitrogen source for aquatic microbial communities is limited, especially in systems with high availability of fixed nitrogen sources, because the use of nitrogen from N_2 for growth ("Nitrogen fixation") requires a high energy investment and only a limited group of organisms (cyanobacteria and some heterotrophic bacteria) is capable of nitrogen fixation Therefore, the main nitrogen sources for microbial communities in coastal waters and sediments are DIN and DON.

Traditionally, studies on aquatic microbial nitrogen uptake focused on uptake of DIN by algae while bacteria were only considered as remineralizers of organic matter. However, the present view of microbial nitrogen uptake is considerably more complex since DON (especially amino acids and urea) has been found to be a potentially important nitrogen source as well and because it is now clear that bacteria are also important players in aquatic microbial nitrogen uptake. In addition to bacteria and algae, aquatic microbial communities can also include Archaea. However, at present their role in total microbial nitrogen uptake is unknown. The pathways for uptake of dissolved nitrogen by algae and bacteria are schematized. The uptake process consists of two steps that are not necessarily coupled. The first step is the uptake of dissolved nitrogenous compounds from the surrounding water into the cells. Small molecules like NH4+, NOx- and urea can easily be taken up into cells while larger compounds like proteins and peptides first need to be broken down into individual amino acids that can either be directly taken up into the cells or be broken down to NH4+ which is then taken up into the cells.

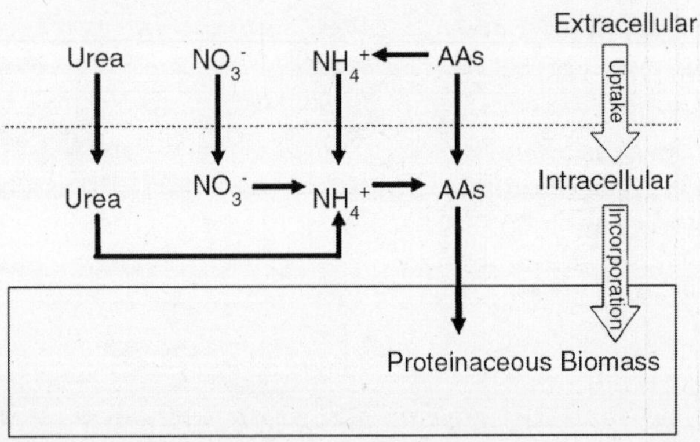

The second step is incorporation of the nitrogen in new biomass. For this, intracellular NO^{3-} and urea are first converted to NH^{4+} which is used to synthesize amino acids that form the building blocks for proteinaceous biomass components. Since amino acids can directly be incorporated into biomass after uptake (*i.e.* little or no conversion is required), these are a highly preferred nitrogen source. Conversely, the microbial preference for NO^{x-} as a nitrogen source is low because of the relatively high energy investment required for its conversion to NH^{4+}. Preferences for NH4+ and urea are generally in between those for amino acids and NO^{x-}.

Fate of Microbial Nitrogen

An interesting, yet poorly studied, aspect of the aquatic nitrogen cycle is the fate of microbial nitrogen (*i.e.*, what happens with the nitrogen after it has been incorporated into microbial biomass?) which is coupled with the fate of other elements in microbial biomass such as carbon.

Although nitrogen can be retained in microbial biomass during continued growth (by cell division), microbial biomass (especially that of bacteria) is often relatively constant over time, meaning that production of new biomass is balanced by loss of biomass due to cell death. Death of microorganisms can be due to various causes such as viral lysis, predation and deteriorating environmental conditions.

After cell death, the leftovers (remnants) of microbial biomass are subject to degradation. During degradation, nitrogen is released back into the system as DON and/or DIN that can be used as nitrogen source for production of new microbial biomass or be lost from the system.

Some microbial biomass components are more resistant to degradation ("Refractory") than others and may accumulate in the system where they serve as a long term sink for microbial-derived nitrogen. Another potential fate of microbial nitrogen is transfer to higher trophic levels of the food web when microbial biomass is consumed by larger organisms.

Stable Isotopes in Ecological and Biogeochemical

Elements such as carbon (C) and nitrogen (N) can occur in different forms called isotopes. Different isotopes of a single element have the same number of protons but a different number of neutrons. Some isotopes are radioisotopes, meaning that these isotopes are subject to radioactive decay. Radioisotopes have been used in ecological and biogeochemical studies (for example for measuring primary production using 14C).

However, work with radioisotopes comes with a potential health risk and can be restricted by law. Moreover, there is no convenient radioisotope of nitrogen which makes that the use of radioisotopes in studies on nitrogen cycling is restricted to work with the short-lived radioisotope 13N (*e.g.*, Zehr *et al.* 1988). Isotopes that are not subject to radioactive decay are stable isotopes. A single element can have two or more stable isotopes, of which one is by far the most abundant (*e.g.* 12C and 14N) while the other forms are rare (for example: 99.6 per cent of all nitrogen in nature is 14N and only 0.4 per cent is 15N). The abundance of these rare isotopes is measured by isotope ratio mass spectrometry (IRMS), a technique that uses a magnetic field to separate gas molecules (CO_2 in case of 13C analysis and N2 in case of 15N analysis) of different molecular weights resulting from different stable isotope compositions.

The relative abundance of these molecules with different molecular weights can be used to determine the relative abundance of the rare (or "Heavy") isotope. This abundance is typically expressed by the delta notation (E) which is the ratio between the rare ("Heavy") isotope and the common (or "Light") isotope relative to that of a standard in permille (per cent°) units.

Table. Carbon and Nitrogen Isotopes

		Type of Isotope	Natural Abundance
Carbon	10_{C}	Radio (½ life 17 Sec)	
	12_{C}	Stable	98.9%
	13_{C}	Stable	1.1%
	13_{C}	Radio (½ life 5730 yr)	
Nitrogen	13_{N}	Radio (½ life 10 min)	
	14_{N}	Stable	99.6%
	15_{N}	Stable	0.4%
	16_{N}	Radio (½ life 7 Sec)	

Stable isotopes can be used in ecological and biogeochemical studies in two different ways. The first way is to study the natural abundance of the rare, heavy isotope. Here, the relevant information lies in small differences in relative abundance of the heavy isotope between different compartments of a system.

These differences are the results of isotopic fractionation, which involves changes in the natural abundance of the heavy isotope during reactions caused by small differences in mass-dependent reaction rates of heavy and light isotopes.

The second way to use stable isotopes is by stable isotope labeling. In this approach, a substrate that is highly enriched in the heavy isotope is deliberately added to a system after which the abundance of the heavy isotope (the "Label") is measured in various components of the system which allows to trace and quantify flows of the label trough these components. The material presented in this thesis all concerns stable isotope labeling work with the nitrogen isotope 15N and the carbon isotope 13C.

BIOMARKERS

Biomarkers are compounds that are unique to a specific group of organisms. Ideally, biomarker compounds are part of the organism's biomass and make up a relatively constant fraction of that biomass so that biomarker concentrations can be converted to total biomass of the group it represents. A variety of compounds have been used as biomarkers for various microbial groups.

Some of these compounds are rapidly degraded after death of the organism, which makes them specific for living biomass ("ecological biomarkers"), for example phospholipid-derived fatty acids (PLFAs). Other components are more resistant to degradation ("biogeochemical biomarkers"), for example D-amino acids from bacterial cell walls, which makes that these components can accumulate in a system, particularly in sediment where the abundance of these biomarkers can provide information on the contribution of material from the corresponding group of organisms to the total organic matter pool in the sediment. A third group of biomarkers are highly refractory molecules ("geological biomarkers").

The concentrations and characteristics of these "molecular fossils" in sediments can be used for paleoenvironmental reconstructions (*e.g.*, Sinninghe Damsté *et al.* 2004). A valuable application of biomarkers in ecological- and biogeochemical studies is analysis of their stable isotope composition, which requires compound-specific stable isotope analysis.

This can be done at natural abundance level or in combination with stable isotope labeling where the latter provides the unique opportunity to directly link microbial identity (biomarker), biomass (biomarker concentration) and activity (label incorporation).

Selection of bacterial biomarkers that have been used in combination with stable isotope analysis. Biomarkers and stable isotope analysis of biomarkers, including some examples of applications in microbial ecology, can be found in Boschker and Middelburg while technical aspects of compound-specific stable isotope analysis.

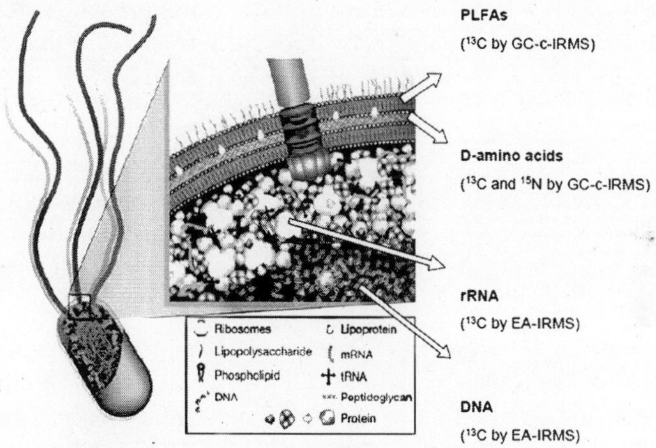

Fig. A Selection of Bacterial Biomarkers used in Combination with Stable Isotope Analysis.

The Problem

Microbial uptake of nitrogen from various nitrogenous substrates is generally studied by analysis of uptake of 15N from 15N-labeled substrates. This approach has frequently been used in water column studies where 15N enrichment of total particulate material in the water is considered to represent total microbial 15N uptake. However, more detailed assessment of 15N uptake by specific microbial groups (bacteria versus algae) has long been hampered by methodological problems. Until recently, two different approaches have been used to discriminate 15N uptake by bacteria versus algae. The first approach is size fractionation, which concerns physical separation of relatively small bacteria from relatively large algae by filtration or flow cytometry.

Although this approach may work in clear waters when a size difference between bacteria and algae is present, size fractionation is not suitable for studies in more turbid waters where bacteria and algae are closely associated with particulate matter. The second approach involves the use of inhibitors to eliminate activity of a specific microbial group, for example antibiotics to inhibit bacterial activity.

However, the use of inhibitors is seriously limited by uncertainties concerning both the specificity and efficiency of inhibitors. Next to the problems in water column studies presented above, studies on microbial nitrogen uptake in sediments have been hampered by even more fundamental methodological problems because of the inability to distinguish between 15N taken up by the microbial community and the remaining added 15N-substrate and/or possible other 15N-labeled extracellular compounds bound to the sediment.

Moreover, another limitation of traditional 15N-labeling work is that it does not allow clean assessment of immobilization of dissolved nitrogen into microbial biomass (the relevant process from a nitrogen cycling perspective as well as from a food web point of view) since total 15N "Uptake" also includes uptake of nitrogenous substrates into microbial cells, which is not necessarily directly coupled to incorporation into biomass. The method presented in this thesis is a new approach that has the potential to overcome the problems mentioned above and therefore may help to further clarify nitrogen flows through bacteria and algae in natural microbial communities.

The Solution

The method presented in this thesis involves the use of compound-specific stable isotope analysis of biomarkers. This approach has already been used at NIOO-CEME for some years in 13C-labeling studies where analysis of 13C incorporation into various group-specific PLFAs provided valuable information on 13C flows through different microbial groups, including bacteria and algae. However, since PLFAs do not contain nitrogen, another type of compounds was required for 15Nlabeling studies.

This thesis investigates the potential of 15N analysis of hydrolysable amino acids (HAAs), including the bacterial biomarker D-alanine (D-Ala). HAAs are amino acids that can be released from proteinaceous biomass (proteins and protein-like material) by hydrolysis in hot acid. HAAs make up a large fraction (50- 80 per cent) of total microbial biomass meaning that analysis of 15N incorporation into total HAAs (THAAs) should provide a good indication of 15N incorporation into total microbial biomass. Amino acids can occur as D- and L-stereoisomers. Most HAAs are L-amino acids that are common biomass constituents of all organisms and therefore have very limited biomarker potential Conversely, hydrolysable D-amino acids are rare and only occur in bacterial biomass, more specifically in peptidoglycan, a structural component of bacterial cell walls.

Therefore, these D-amino acids can be used as bacterial biomarkers. Although different D-amino acids can be present in peptidoglycan, our method focuses on D-Ala because this is the only D-amino acid that is present in all bacteria and because it makes up a relatively stable fraction of total bacterial biomass. Moreover, D-Ala has been shown to be most suitable for stable isotope analysis by GC-c-IRMS.

The method can also be used to measure incorporation of the carbon isotope 13C into D-Ala and other HAAs, which allows assessment of carbon flows through bacteria and algae. This is particularly interesting in direct combination with analysis of nitrogen flows..

MICROBES AND AIR POLLUTION

This is an era of increased awareness of how society has an impact on the environment. Due to the demands by the public, government has placed huge burdens on industry to operate in a more environmentally responsible manner. To meet this challenge, industry has had to turn to a variety of fields to meet these tougher environmental standards. To help eliminate environmental hazards, industries employ physical, chemical, thermal, and biological processes. This paper will discuss biotechniques used to control air pollution and examine how a Hyphomicrobium bacteria strain may be used to filter volatile sulfur compounds from industrial emissions.

Over the last fifteen years biological treatment technologies have gained increased popularity due to their low costs, operational simplicity, and because they are cleaner than alternative technologies. Industry especially like them because there is minimal requirement of energy and raw materials. Also there is a minimal production of waste materials to be disposed of. These biotechniques include biofiltration, bioscrubbers, and biotrickling filters. I will limit discussion to biofiltration because it is the most common method employed and it is the simplest to explain.

A biofilter consists of a solid phase, consisting of a porous media, on which microorganisms are allowed to grow. A stream of gas containing unwanted molecules is then passed through the solid phase. The resulting microbial action leads to the destruction of the contaminants. The solid phase of the biofilter may consist of soil, organic matter, activated carbon, clay particles or a porous glass. When the volatile compound is first passed through the filter it is sorbed by the solid phase. Then the sorbed molecule is metabolized by the film of microorganisms adhering to the solid phase. The microorganism uses the unwanted pollutant as a carbon and energy source and this results in changing a pollutant into an inert substance, such as carbondioxide and water.

Volatile sulfur compounds have caused serious environmental problems due to their high odor index and toxicity. It has also been shown that sulfur containing compounds can be oxidized in the troposphere causing acid rain and contributes to cloud formation. Volatile sulfur compounds can be formed in various microbial ecosystems such as compost stacks and sewage treatment plants. Industrial processes such as brewing and paper manufacture, both important industries in Wisconsin, are also major contributors to this problem.

Research is now being done to increase our knowledge of the biodegration of these complex volatile compounds. Researchers are revealing new microbial strains that contain metabolic pathways that can degrade these compounds into inert substances. These microorganisms can then be inoculated into the solid phase of the biofilter where they can adhere to and grow using the compounds passed through the filter as a carbon and energy source. Researchers at the University of Nijmegen in the Netherlands have isolated and characterized a strain of Hyphomicrobium bacteria that is able to use dimethylsulfide, a major contributor to the volatile sulfur compounds released in air pollution emissions, as a carbon and energy source.

Hyphomicrobium cells have been successfully isolated and enriched in batch cultures of ethanol and methanol. This strain of Hyphomicrobium was designated strain VS. The cells were ovoid with a single polar tubular filament. The filaments were up to three times as long as the length of the cell and it produced buds. Motility was observed for small cells and maturing buds. Hyphomicrobium VS were unable to grow on complex media or under anaerobic conditions.

Experiments were done to demonstrate whether or not Hyphomicrobium VS could be applied for use in biofiltration. Two different filters were constructed with different solid phases. The first one was packed with polyurethane. The second one was packed with lava stone. Each was inoculated with a methanol grown culture of Hyphomicrobium VS. Starting at a low concentration of dimethylsulfide and slowly increasing the concentration over the forty-five days of the experiment researchers attempted to find the maximum operating capacity of the filters. The polyurethane filter reached a capacity of 5.3 mmol per liter per day. The lava stone filter did not reach its maximum operating capacity and the removal capacity of this filter had increased during the course of this experiment. This demonstrates how the type of solid phase within the filter can have a dramatic effect on the filter's performance.

These experiments have shown Hyphomicrobium to be a promising organism for application in biofiltration of air containing dimethylsulfide. It showed that this strain could be easily be cultured on a large scale on methanol and only require a short amount of time to adapt to dimethylsulfide. During the 45 day period of these experiments it demonstrated up to a 97 per cent efficiency at removing the sulfur compounds at high air velocities.

Hyphomicrobium has an extra advantage for odor treatment that it also can degrade methylated amines.

The future of this as a practical solution to reduce sulfur emissions rests with more research. Further studies will have to be done that will focus on the use of different packing materials in order to maximize efficiency and to increase the capacity of the filter. Both polyurethane and lava stone appeared to be good package materials but it was shown that the type of material chosen may have a significant effect on the maximum removal rates of volatile compounds. Also there must be a study of the log term operation of this type of filter at sulfur concentrations equal to those found in industrial exhaust gases before any final conclusions can be drawn.

There are several alternatives to the use of biofilters to reduce air pollution. These include: centrifugation with gas cyclones, adsorption with activated carbon, filtration with canvas filters, and catalytic incineration. Some of the advantages these physical and chemical techniques have compared to biotechniques are: a more complete elimination of particles, good reliability, and the possibility of recovering these compounds to recycle them into other products. The disadvantages these physical and chemical processes have include: high investment costs in material, high operational costs, the production of waste products that may also need treatment facilities, and final disposal of waste products.

I feel that biotechnology provides some excellent opportunities for the future of pollution controls. If the microorganisms can be found that can biodegrade the toxic compounds produced by industry it would take society along way towards solving many of the problems we face with the environment. These techniques may also be used in conjunction with physical or chemical process to increase their effectiveness. This type of technology could help eliminate pollution and recover already contaminated industrial areas. If nature can provide solutions to many of the types of pollutants we face, it could help industry coexist with the environment.

There is a growing interest in biotechnology as a method of reducing air pollution. People around the world are realizing that these techniques are an efficient cost effective way to reduce pollution. The application of this technology is still expanding so higher flow rates and higher concentrations can be treated. So far implementation of these techniques lags behind of the potential it offers to society. Biotechniques could play an important role in controlling air pollution in the future if its full potential is realized.

A biofilter which eliminated ethylene (C_2H_4) from the high parts-per-million range to levels near the limit for plant hormonal activity (0.01 to 0.1 ppm) was developed. Isolated ethylene-oxidizing bacteria were immobilized on peat-soil in a biofilter (687 cm3) and subjected to an atmospheric gas flow (73.3 ml min"1) with 2 or 117 ppm of C_2H_4. Ethylene was eliminated to a minimum level of 0.017 ppm after operation with 2.05 ppm of C_2H_4 for 16 days. Also, the inlet C_2H_4 concentration of 117 ppm was reduced to <0.04 ppm. During operation with 2 and 117 ppm of C_2H_4, an increase in the C_2H_4 removal rate was observed, which was attributed to proliferation of the immobilized bacteria, notably in the first 0- to 5-cm segment of the biofilter. The maximal C_2H_4 elimination capacity of the biofilter was 21 g of C_2H_4 m"3 day"1 during operation with 117 ppm of C_2H_4 in the inlet gas. However, for the first 0- to 5-cm segment of the biofilter, an elimination capacity of 146 g of C_2H_4 m"3 day"1 was calculated. Transition of the biofilter temperature from 21 to 10°C caused a 1.6-fold reduction in the C_2H_4 removal rate, which was reversed during operation for 18 days. Batch experiments with inoculated peat-soil demonstrated that C_2H_4 removal still occurred after storage at 2, 8, and 20°C for 2, 3, and 4 weeks. However, the C_2H_4 removal rate decreased with increasing storage time and was reduced by ca. 50 per cent after storage for 2 weeks at all three temperatures. The biofilter could be a suitable tool for C_2H_4 removal in, *e.g.*, horticultural storage facilities, since it (i) removed C_2H_4 to 0.017 ppm, (ii) had a good operational stability, and (iii) operated efficiently at 10°C.

Ethylene (C_2H_4) is a gaseous plant hormone and air pollutant which has a strong effect on plant physiological processes such as ripening, senescence, and aging. The threshold limit for C_2H_4 sensitivity varies among plant species, but a general dose response is no effect below 0.01 ppm, half-maximal effect at 0.1 ppm, and maximal effect at 1 to 10 ppm. The lower threshold limit of 0.01 ppm is only slightly higher than the atmospheric background concentration (0.003 to 0.005 ppm) and may be exceeded in polluted areas. Also, significant accumulation of C_2H_4 may occur in horticultural storage facilities due to endogenous production by the plant material.

To reduce the detrimental effects of such C_2H_4, chemical C_2H_4 scrubbers are widely used in storage facilities for horticultural produce. A drawback of such scrubbers is the cost of operation and the need for replenishment of the C_2H_4-removing agents. Therefore, the use of biological catalysts for C_2H_4 removal is an interesting alternative in horticulture. Likewise, biological catalysts could be used to reduce high ethylene concentrations at point sources, such as petrochemical facilities, where ethylene and polyethylene are produced.

Previously, biological waste gas purification has been reported for several air pollutants, including C_2H_4. So far, however, no biological system for C_2H_4 removal which has a satisfactory operational stability and a sufficient efficiency to reduce the C_2H_4 concentration to levels near the lower threshold limit for the plant hormonal response (0.01 ppm) has been described.

Here I report the outcome of a study with a biofilter based on isolated ethylene-oxidizing bacteria that were immobilized on peat-soil. During stable operation with inlet levels of 2 and 117 ppm of C_2H_4, it was possible to reduce the outlet levels to less than 0.04 ppm with a minimum level of 0.017 ppm.

The ethylene-oxidizing bacterial strain RD-4 was isolated and cultivated as described by Elsgaard and Anderse. For the present purpose, the strain was grown in 5-liter vessels containing 2 liters of defined mineral medium without added C sources. The medium in each bottle was inoculated with 20 ml of a pregrown culture, and C_2H_4 was added to a headspace concentration of ca. 1 per cent. The cultures were incubated with shaking (120 rpm) at 30°C, and the C_2H_4 concentration was replenished to ca. 1 per cent when it dropped below 0.5 per cent (assayed by gas chromatographic [GC] analysis). After incubation for 5 days, the ethylene-oxidizing bacteria were harvested by centrifugation (16.000 × *g* for 15 min) and resuspended in 1.5 liters of autoclaved tap water. The density of the cell suspension was estimated to be 2×10^8 cells ml^{-1} by direct microscopic counts in a Bürker-Türk counting chamber. The cell suspension was stored at 0 to 2°C until needed for biofilter experiments (1 week later) and batch experiments (2 weeks later).

Biofilter construction. A laboratory-scale biofilter was made from an acrylic core with an inner diameter of 5 cm and a length of 40 cm. Six butyl rubber stoppers (for gas sampling) were inserted in holes at a vertical distance of 5 cm apart.

The core was stoppered with rubber stoppers, each equipped with a T-connector. The T-connectors served as channels for the gas flow and had a butyl rubber stopper inserted for gas sampling. In operational mode, the biofilter had a volume of 687 cm^3 and could be sampled at depths from 0 to 35 cm (inlet and outlet, respectively) at increments of 5 cm.

A mixture of atmospheric air and C_2H_4 in N_2 (AGA SpecialGas, Lidingö, Sweden) was supplied to the biofilter by use of two mass flow controllers, with a resulting flow rate (*F*) of 73 ml min^{-1}. The pseudo-residence time ($è_p = V/F$) of the biofilter was 9.4 min. Before reaching the inlet, the gas was humidified by being bubbled through a flask containing distilled water. To verify the gas flow was constant during the experiments, a digital flowmeter (Jour Research, Onsala, Sweden) was connected to the outlet of the biofilter and the readings of the flowmeter were checked on a daily to weekly basis.

Immobilized ethylene-oxidizing bacteria were prepared by mixing 300 g of peat-soil (Pindstrup blend 2; Pindstrup Mosebrug) with 200 ml of distilled water and 100 ml of bacterial cell suspension. The soil sample was thoroughly mixed and allowed to equilibrate at room temperature (20 to 22°C) for 1 h. The inoculated peat-soil was loosely packed in the biofilter to a density of 0.44 g cm^{-3}.

The final dry-matter content of the inoculated peat-soil was 20.5 per cent.

- Operation with 2 ppm of C_2H_4. A mixture of 10 ppm of C_2H_4 and atmospheric air (1:4) was applied to the biofilter. During an operational period of 16 days at 20 to 22°C, gas samples (0.6 ml) were collected from all depths of the biofilter and analyzed for C_2H_4.

Gas samples (0.2 to 0.5 ml) for measurement of O_2 and CO_2 at the biofilter inlet and outlet were collected after 1 day of operation.

- Operation with different C_2H_4 concentrations. After 16 days of operation with 2 ppm of C_2H_4, the biofilter was operated with atmospheric air for 25 days (*i.e.*, the biofilter was starved for C_2H_4). Subsequently, the inlet level was adjusted to 2 ppm of C_2H_4 for 8 days and then to 117 ppm for 80 days. The latter concentration was obtained by mixing 1,000 ppm of C_2H_4 and atmospheric air (1:8). Gas samples for C_2H_4 analysis were withdrawn regularly.
- Transition from 21 to 10°C. During operation with 117 ppm of C_2H_4, the influence of temperature was tested by placing the biofilter at 10°C in a thermostated incubator. Control experiments showed that temperature equilibrium at 10°C was reached in the center of the biofilter within 5 h. Ethylene concentrations were measured at the biofilter inlet and outlet during operation for 18 days.

Batch experiments were done to test the effect of storage on the C_2H_4 removal rate of the inoculated peat-soil. A sample of immobilized bacteria was prepared as described above, while a control sample was prepared by mixing 100 g of peat-soil with 100 ml of distilled water. Each of the soil samples was divided into three portions that were placed at 2, 8, and 20°C. To test the C_2H_4 removal rate before storage, triplicate 10-g soil samples were transferred to 120-ml serum bottles that were purged with atmospheric air and closed with butyl rubber stoppers. Ethylene was added to a headspace concentration of ca. 20 ppm, and the time course of C_2H_4 removal (20 to 22°C) was monitored by GC analysis of withdrawn gas samples (0.4 ml). Similar assays were done with the stored soil samples after 2, 3, and 4 weeks of storage. Before these assays, the 10-g subsamples were allowed to equilibrate at room temperature for 1 h.

Gas samples for analysis of C_2H_4, CO_2, and O_2 were collected with 1-ml gas-tight syringes (Pressure-Lok). Generally, C_2H_4 was quantified with a Hewlett-Packard 5840A gas chromatograph equipped with a flame ionization detector. For injection of a 0.6-ml gas sample, the C_2H_4 detection limit was 0.04 ppm for a signal-to-noise ratio of 3.

Occasionally, samples from the biofilter were collected by passing the outlet flow through a stoppered 120-ml serum bottle for 1 to 2 h with two hypodermic needles. The bottle was then removed from the gas stream, and subsamples of 1.0 ml were analyzed for C_2H_4 with a Photovac 10Splus gas chromatograph equipped with a photoionization detector. The ethylene detection limit was 0.002 ppm for a sample volume of 1.0 ml.

Oxygen (0.5-ml samples) was measured on a Varian 3700 gas chromatograph with a thermal conductivity detector (TCD) and a molecular sieve 5A column. The oven temperature was 30°C, and the carrier gas was He (flow rate, 76 ml min^{-1}). Carbon dioxide (0.2-ml samples) was measured on a GC 82 (Mikrolab, Højbjerg, Denmark) with a TCD. The column (Porapak N) was operated at 60°C with He as the carrier gas (flow rate, 43 ml min^{-1}).

Statistics. Data from replicate samples are presented as mean ± standard error (SE) with the number of samples (n) indicated. The significance of differences between data from the biofilter inlet and outlet was tested by use of a two-tailed Student t test.

During the experiments, the flow rate measured at the outlet of the biofilter ranged from 71.4 to 75.9 ml min^{-1} with a mean and SE of 73.3 ± 0.3 ml min^{-1} (n = 23). In a control experiment, it was demonstrated that the flow rates at the inlet and the outlet were identical (76.5 to 76.8 ml min^{-1}) and equaled the sum of the flow rates for the two gaseous components that were mixed (67.9 and 8.7 ml min^{-1}).

Furthermore, it was demonstrated that with an empty biofilter, the outlet C_2H_4 concentration (2.19 ± 0.03 ppm; n = 3) was equal to the inlet concentration (2.18 ± 0.03 ppm; n = 3). This showed that no significant leakage or adsorption occurred.

- Operation with 2 ppm of C_2H_4. During the first experiment, a stable inlet concentration of 2.05 ± 0.01 ppm (n = 57) was applied to the biofilter. Measurements of the outlet C_2H_4 concentration after 1 h of operation (0.23 ± 0.01 ppm; n = 4) demonstrated that 89 per cent of the incoming C_2H_4 had already been removed by the biofilter at this early stage. During the experiment, the efficiency of C_2H_4 removal gradually increased to 99 per cent. Thus, at the end of the experiment the biofilter had a stable performance with an outlet concentration of only 0.017 to 0.020 ppm of C_2H_4.

Gas analysis at different biofilter soil depths showed that after 1 h of operation all the soil layers were exposed to C_2H_4 at concentrations ranging from 2.05 to 0.23 ppm. However, as the rate of C_2H_4 removal increased in the soil layers close to the inlet, removal of the incoming C_2H_4 ultimately occurred within the first 15 cm of the biofilter. Segment-specific C_2H_4 removal rates were calculated as follows:

$$SR(s) = \Delta C_2H_4(s) \times F \times \rho(C_2H_4) \times M(s)^{-1}$$

where SR(s) is the specific C_2H_4 removal rate (micrograms of C_2H_4 per gram [dry weight] of soil per hour) by a given 5-cm segment (s), $\Delta C_2H_4(s)$ is the difference between the inlet and outlet C_2H_4 concentration of the segment (microliters per liter), F is the flow rate (4.4 liters h^{-1}), ρ (C_2H_4) is the density of C_2H_4 at 21°C (1.16 μg μl^{-1}), and M(s) is the soil dry-matter content in a 5-cm segment (8.8 g). These calculations showed that the initial C_2H_4 removal rate in each segment (after 1 h of operation) depended linearly on the C_2H_4 concentration (data not shown). In the 0- to 5-cm segment, the C_2H_4 removal rate increased from 0.35 to 0.89 ìg of C_2H_4 g (dry weight) of $soil^{-1}$ h^{-1} during the experiment. In the subsequent segments, a relatively stable or decreasing rate was observed.

Oxygen measurements after 1 day of operation showed that the O_2 content of the inlet gas and the outlet gas was 15.6 per cent ± 0.1 per cent ($n = 3$) and 15.2 per cent ± 0.1 per cent ($n = 3$), respectively. This demonstrated that oxic conditions prevailed throughout the whole soil column. At the same time, the CO_2 concentrations at the inlet and the outlet were 0.018 per cent ± 0.001 per cent ($n = 4$) and 0.024 per cent ± 0.001 per cent ($n = 3$), respectively. Thus, the outlet CO_2 concentration was significantly higher than the inlet CO_2 concentration ($P < 10^{-4}$), demonstrating that a net mineralization of organic C occurred in the biofilter.

- Operation with different C_2H_4 concentrations. Operation for 25 days without C_2H_4 in the inlet gas caused a decrease in the biofilter efficiency of C_2H_4 removal. When 2 ppm C_2H_4 was reapplied, only 11 to 15 per cent of the incoming C_2H_4 was removed after 0.5 to 3 h of operation; however, after operation for 8 days, the efficiency of C_2H_4 removal was recovered and >98 per cent of the incoming C_2H_4 was removed by the biofilter. Transition of the inlet C_2H_4 level from 2 to 117 ppm similarly caused a transient decrease in the efficiency of the biofilter, which initially (after 0.5 h) removed only 10 per cent of the incoming C_2H_4 (after the transition, the inlet concentration was 117.2 ± 0.4 ppm [$n = 76$] during the subsequent experiments). However, after operation at 117 ppm of C_2H_4 for 4 days, the C_2H_4 concentration at the outlet of the biofilter was below the detection limit of 0.04 ppm. Thus, more than 99.9 per cent of the incoming C_2H_4 was removed by the biofilter.

The long-term operational stability of the biofilter was tested with the inlet level of 117 ppm of C_2H_4. It was demonstrated that for more than 75 days of constant operation, the biofilter was able to reduce the outlet C_2H_4 concentration to less than 0.04 ppm. Most of the C_2H_4 was removed during passage through the first 0 to 5 cm of the biofilter. However, after 70 days of operation, the C_2H_4 removal by the 0- to 5-cm segment started to decrease but the removal was then accomplished within the 5- to 10-cm segment.

- Transition from 21 to 10°C. When the biofilter was transferred from 21 to 10°C, the outlet C_2H_4 concentration gradually increased from <0.04 to 46.6 ± 0.3 ppm ($n = 3$). However, after 2 days of operation at 10°C, the outlet concentration started to decrease with a linear time course. Thus, when the experiment was stopped after 18 days, the outlet concentration was 1.6 ± 0.1 ppm C_2H_4 ($n = 3$), which was equivalent to a removal efficiency of 98.6 per cent.

With freshly inoculated peat-soil, rapid depletion of the added 20 ppm of C_2H_4 was observed During the first 1 h of the experiment, C_2H_4 was removed at rate of 17.1 ppm h^{-1}, which was equal to a specific rate of 1.06 ìg of C_2H_4 g (dry weight) of soil^{-1} h^{-1}. After storage of the inoculated peat-soil for 28 days at 20°C, a constitutive C_2H_4 removal still occurred but took place at a lower rate than for the freshly inoculated soil, so that the data for the first 1.5 h of the experiment showed a removal rate of 3.3 ppm h^{-1} (0.21 ìg of C_2H_4 g [dry weight] of $soil^{-1}$ h^{-1}). Within the assay time of 5 h, no C_2H_4 removal occurred in samples of uninoculated peat-soil either immediately or after 28 days of storage at 20°C.

All time courses of C_2H_4 removal by inoculated peat-soil stored at 2, 8, or 20°C for 0, 2, 3, and 4 weeks fell within the range depicted by the data from day 0 and day 28 in. Thus, for comparison, the initial C_2H_4 removal rate for each assay was calculated from the first two or three data points obtained during 0 to 1.5 h of incubation. These rates demonstrated that the C_2H_4 removal rate decreased as a result of longer storage times. However, the extent of the decrease caused by the storage time was almost unaffected by the storage temperature.

8

Environmental Pollution: Control and Prevention

Over the course of the twentieth century, growing recognition of the environmental and public health impacts associated with anthropogenic activities has prompted the development and application of methods and technologies to reduce the effects of pollution. In this context, governments have adopted regulatory and other policy measures to minimize negative effects and ensure that environmental quality standards are achieved. The objective of this chapter is to provide an orientation to the methods that are applied to control and prevent environmental pollution.

The basic principles followed for eliminating negative impacts on the quality of water, air or land will be introduced; the shifting emphasis from control to prevention will be considered; and the limitations of building solutions for individual environmental media will be examined. It is not enough, for example, to protect air by removing trace metals from a flue gas only to transfer these contaminants to land through improper solid waste management practices. Integrated multimedia solutions are required.

APPROACH OF THE POLLUTION CONTROL

The environmental consequences of rapid industrialization have resulted in countless incidents of land, air and water resources sites being contaminated with toxic materials and other pollutants, threatening humans and ecosystems with serious health risks.

More extensive and intensive use of materials and energy has created cumulative pressures on the quality of local, regional and global ecosystems. Before there was a concerted effort to restrict the impact of pollution, environmental management extended little beyond laissez-faire tolerance, tempered by disposal of wastes to avoid disruptive local nuisance conceived of in a short-term perspective.

The need for remediation was recognized, by exception, in instances where damage was determined to be unacceptable. As the pace of industrial activity intensified and the understanding of cumulative effects grew, a *pollution control* paradigm became the dominant approach to environmental management.

Two specific concepts served as the basis for the control approach:

- The *assimilative capacity* concept, which asserts the existence of a specified level of emissions into the environment which does not lead to unacceptable environmental or human health effects
- The *principle of control* concept, which assumes that environmental damage can be avoided by controlling the manner, time and rate at which pollutants enter the environment

Under the pollution control approach, attempts to protect the environment have especially relied on isolating contaminants from the environment and using end-of-pipe filters and scrubbers. These solutions have tended to focus on media-specific environmental quality objectives or emission limits, and have been primarily directed at point source discharges into specific environmental media (air, water, soil).

APPLYING POLLUTION CONTROL TECHNOLOGIES

Application of pollution control methods has demonstrated considerable effectiveness in controlling pollution problems - particularly those of a local character. Application of appropriate technologies is based on a systematic analysis of the source and nature of the emission or discharge in question, of its interaction with the ecosystem and the ambient pollution problem to be addressed, and the development of appropriate technologies to mitigate and monitor pollution impacts.

In their article on air pollution control, Dietrich Schwela and Berenice Goelzer explain the importance and implications of taking a comprehensive approach to assessment and control of point sources and non-point sources of air pollution. They also highlight the challenges - and opportunities - that are being addressed in countries that are undergoing rapid industrialization without having had a strong pollution control component accompanying earlier development. Marion Wichman-Fiebig explains the methods that are applied to model air pollutant dispersion to determine and characterize the nature of pollution problems. This forms the basis for understanding the controls that are to be put into effect and for evaluating their effectiveness. As the understanding of potential impacts has deepened, appreciation of effects has expanded from the local to the regional to the global scale.

Hans-Ulrich Pfeffer and Peter Bruckmann provide an introduction to the equipment and methods that are used to monitor air quality so that potential pollution problems can be assessed and the effectiveness of control and prevention interventions can be evaluated.

John Elias provides an overview of the types of air pollution controls that can be applied and the issues that must be addressed in selecting appropriate pollution control management options.

The challenge of water pollution control is addressed by Herbert Preul in an article which explains the basis whereby the earth's natural waters may become polluted from point, non-point and intermittent sources; the basis for regulating water pollution; and the different criteria that can be applied in determining control programmes. Preul explains the manner in which discharges are received in water bodies, and may be analysed and evaluated to assess and manage risks. Finally, an overview is provided of the techniques that are applied for large-scale wastewater treatment and water pollution control.

A case study provides a vivid example of how wastewater can be reused - a topic of considerable significance in the search for ways that environmental resources can be used effectively, especially in circumstances of scarcity. Alexander Donagi provides a summary of the approach that has been pursued for the treatment and groundwater recharge of municipal wastewater for a population of 1.5 million in Israel.

COMPREHENSIVE WASTE MANAGEMENT

Under the pollution control perspective, waste is regarded as an undesirable by-product of the production process which is to be contained so as to ensure that soil, water and air resources are not contaminated beyond levels deemed to be acceptable. Lucien Maystre provides an overview of the issues that must be addressed in managing waste, providing a conceptual link to the increasingly important roles of recycling and pollution prevention.

In response to extensive evidence of the serious contamination associated with unrestricted management of waste, governments have established standards for acceptable practices for collection, handling and disposal to ensure environmental protection. Particular attention has been paid to the criteria for environmentally safe disposal through sanitary landfills, incineration and hazardous-waste treatment.

To avoid the potential environmental burden and costs associated with the disposal of waste and promote a more thorough stewardship of scarce resources, waste minimization and recycling have received growing attention. Niels Hahn and Poul Lauridsen provide a summary of the issues that are addressed in pursuing recycling as a preferred waste management strategy, and consider the potential worker exposure implications of this.

SHIFTING EMPHASIS TO POLLUTION PREVENTION

End-of-pipe abatement risks transferring pollution from one medium to another, where it may either cause equally serious environmental problems, or even end up as an indirect source of pollution to the same medium. While not as expensive as remediation, end-of-pipe abatement can contribute significantly to the costs of production processes without contributing any value. It also typically is associated with regulatory regimes which add other sets of costs associated with enforcing compliance. While the pollution control approach has achieved considerable success in producing short-term improvements for local pollution problems, it has been less effective in addressing cumulative problems that are increasingly recognized on regional (*e.g.*, acid rain) or global (*e.g.*, ozone depletion) levels.

The aim of a health-oriented environmental pollution control programme is to promote a better quality of life by reducing pollution to the lowest level possible. Environmental pollution control programmes and policies, whose implications and priorities vary from country to country, cover all aspects of pollution (air, water, land and so on) and involve coordination among areas such as industrial development, city planning, water resources development and transportation policies.

Thomas Tseng, Victor Shantora and Ian Smith provide a case study example of the multimedia impact that pollution has had on a vulnerable ecosystem subjected to many stresses - the North American Great Lakes. The limited effectiveness of the pollution control model in dealing with persistent toxins that dissipate through the environment is particularly examined. By focusing on the approach being pursued in one country and the implications that this has for international action, the implications for actions that address prevention as well as control are illustrated.

As environmental pollution control technologies have become more sophisticated and more expensive, there has been a growing interest in ways to incorporate prevention in the design of industrial processes - with the objective of eliminating harmful environmental effects while promoting the competitiveness of industries. Among the benefits of pollution prevention approaches, clean technologies and toxic use reduction is the potential for eliminating worker exposure to health risks.

David Bennett provides an overview of why pollution prevention is emerging as a preferred strategy and how it relates to other environmental management methods. This approach is central to implementing the shift to sustainable development which has been widely endorsed since the release of the United Nations Commission on Trade and Development and reiterated at the Rio United Nations Conference on Environment and Development (UNCED) Conference. The pollution prevention approach focuses directly on the use of processes, practices, materials and energy that avoid or minimize the creation of pollutants and wastes at source, and not on "add-on" abatement measures.

While corporate commitment plays a critical role in the decision to pursue pollution prevention, Bennett draws attention to the societal benefits in reducing risks to ecosystem and human health-and the health of workers in particular. He identifies principles that can be usefully applied in assessing opportunities for pursuing this approach.

IMPACT OF MAN'S ACTIVITIES IN POLLUTANTS

The addition of certain pollutants like the carbon dioxide gases increases the temperature of the earth. These gases remain close to the surface of earth, forming an insulating layer. The surface becomes hot due to the solar radiations.

This radiation is normally reflected back into the space. However, due to the presence of the insulating layer the heat remains within the earth's atmosphere and increases its temperature. The reduction in the forest cover also contributes to the heating of earth. This is called global warming.

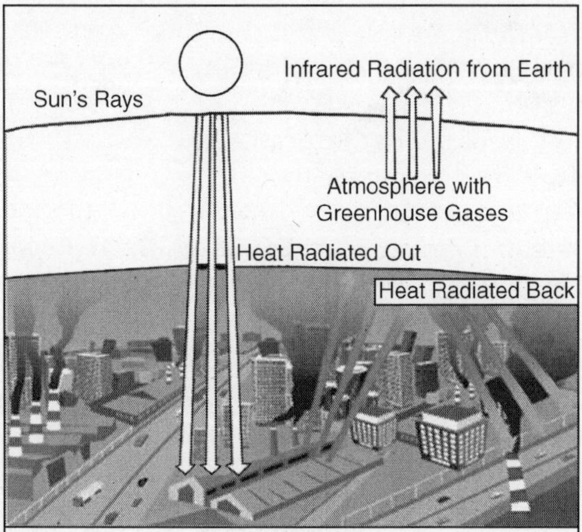

Global Warming

Global warming results in the melting of polar ice caps. This inturn results in the rise in the level of ocean waters. It is also called the greenhouse effect. Greenhouses are the covered areas where the temperature is artificially kept higher than the surrounding cold environment for the growth of plan

OZONE DEPLETION

Ozone (O_3) IS a molecule formed BY three atoms of oxygen WHILE oxygen (O_2), IS essential FOR ALL aerobic forms of life. Ozone IS a deadly poison. However, AT the higher levels of the atmosphere, ozone performs an essential function. It shields the surface of the earth FROM ultraviolet (UV) radiation FROM the Sun. UV radiation causes skin cancer, damage TO eyes, immune system. It can also affect GLOBAL rainfall AND cause ecological disturbances.

Ozone AT the higher levels of the atmosphere IS a product of UV radiation acting ON oxygen (O_2) molecule. The higher energy UV radiations split apart SOME molecular oxygen (O_2) INTO free oxygen (O) atoms. These atoms then combine with the molecular oxygen to form ozone AS shown IN the equations below.

$$O_2 \xrightarrow{UV} O + O$$

$$O + O_2 \longrightarrow \underset{\text{ozone}}{O_3}$$

The decline of ozone content IN the atmosphere began IN the 1980s. This decrease has been linked TO synthetic chemicals LIKE chlorofluorocarbons (CFCs) which are used AS refrigerants AND IN fire extinguishers. IN 1987, the United Nations Environment Programme (UNEP) succeeded IN forging an agreement TO freeze CFC production.

BIOLOGICAL MAGNIFICATION

Extensive usage of several pesticides and other chemicals to protect our crops from diseases and pests has resulted in the entry of these harmful chemicals into our bodies through the food chain. These chemicals are either washed down into the soil or into the water bodies. From the soil, these are absorbed by the plants along with water and minerals, and from the water bodies these are taken up by aquatic plants and animals. This is one of the ways in which they enter the food chain. As these chemicals are not degradable, they get accumulated progressively at each trophic level.

As human beings occupy the top level in any food chain, the maximum concentration of these chemicals gets accumulated in our bodies. This phenomenon is known as biological magnification.

This is the reason why our food grains such as wheat and rice, vegetables and fruits, and even meat, contain varying amounts of pesticide residues. They cannot always be removed by washing or other means.

DID U KNOW

Persistent organic pollutants [pop] or pesticides enter our food chain through various methods. DDT, a very harmful pesticide has been banned from use. Almost all ready made foods contain two or more pesticides. Among fruits and vegetables peaches, apples, spinach and lettuce have the highest amount of pesticide content. Onions have the least followed by papaya. Dieldrin and DDE are the most commonly used pesticides. DDE is a break down product of DDT. Dieldrin is an organochlorine pesticide. Use of organically grown food items is one way of reducing pesticide intake.

HUMAN IMPACT ON THE NATURAL ENVIRONMENT

Natural environment is of crucial importance for social and economic life.

We use the living world as:

- A resource for food supply
- An energy source
- A source for recreation

- A major source of medicines
- Natural resources for industrial products

In this respect the diversity of nature not only offers man a vast power of choice for his current needs and desires. It also enhances the role of nature as a source of solutions for the future needs and challenges of mankind.

STATE OF ECOSYSTEMS, HABITATS AND SPECIES

In the past, human interaction with nature, although often having a disruptive effect on nature, often also enriched the quality and variety of the living world and its habitats - *e.g.* through the creation of artificial landscapes and soil cultivation by local farmers. Today, however, human pressure on natural environments is greater than before in terms of magnitude and efficiency in disrupting nature and natural landscapes, most notably:

- Intensive agriculture replacing traditional farming; this combined with the subsidies of industrial farming has had an enormous effect on western rural landscapes and continues to be a threat.
- Mass tourism affecting mountains and coasts.
- The policies pursued in the industry, transport and energy sectors having a direct and damaging impact on the coasts, major rivers (dam construction and associated canal building) and mountain landscapes.
- The strong focus of forestry management on economic targets primarily causes the decline in biodiversity, soil erosion and other related effects.

The clearest manifestations of the degradation of the natural environment are:

Reduction and Fragmentation of Habitats and Landscapes

The expansion of humans activities into the natural environment, manifested by urbanisation, recreation, industrialisation, and agriculture, results in increasing uniformity in landscapes and consequential reduction, disappearance, fragmentation or isolation of habitats and landscapes.

It is evident that the increasing exploitation of land for human use greatly reduces the area of each wildlife habitat as well as the total area surface throughout Europe.

The consequences are:

- A decreased species diversity, due to reduced habitable surface area which corresponds to a reduced "species carrying capacity".
- The reduction of the size of habitats also reduces the genetic diversity of the species living there. Smaller habitats can only accommodate smaller populations, this results in an impoverished gene pool.
- The reduction of genetic resources of a species diminishes its flexibility and evolutionary adaptability to changing situations. This has significant negative impacts on its survival.

The conditions under which the reduction of habitats often occur prevent living organisms making use of their normal ways to flee their threatened habitat. Those escape routes include migration to other habitats, adaption to the changing environment, or genetic interchange with populations in nearby habitats.

Of particular concern is:

- The abrupt nature of human intervention; human projects are planned and implemented on a much shorter time scale than natural processes;
- Furthermore human intervention, such as the construction of buildings, motorways or railways results in the fragmentation of habitats, which strongly limits the possibility for contact or migration among them;
- In extreme cases even the smallest, narrowest connections between habitats are broken off. Such isolation is catastrophic for life in the habitat fragments.

Loss of Species of Fauna and Flora

Although relatively few species of Europe's fauna and flora have actually become totally extinct during this century, the continent's biodiversity is affected by decreasing species numbers and the loss of habitats in many regions. Approximately 30 per cent of the vertebrates and 20 per cent of the higher plants are classified as "threatened". Threats are directly linked to the loss of habitats due to destruction, modification and fragmentation of ecosystems as well as from overuse of pesticides and herbicides, intensive farming methods, hunting and general human disturbance. The overall deterioration of Europe's air and water quality add to the detrimental influence.

AGRICULTURE

Europe's natural environment is inextricably linked with agriculture and forestry. Since agriculture traditionally depends on sound environmental conditions, farmers have a special interest in the maintenance of natural resources and for centuries maintained a mosaic of landscapes which protected and enriched the natural environment. As a result of needs for food production since the 1940s, policies have encouraged increased pro- duction through a variety of mechanisms, including price support, other subsidies and support for research and development. The success achieved in agricultural production has however entailed increased impact on the environment.

Modern agriculture is responsible for the loss of much wildlife and their habitats in Europe, through reduction and fragmentation of habitats and wildlife populations. The drainage of wetlands, the destruction of hedgerows and the intensive use of fertilizers and pesticides can all pose a threat to wildlife. Highly specialised monoculture are causing significant loss in species abundance and diversity. On the other hand increased production per hectare in intensive areas, raising of livestock volume, and lower prices for agricultural products also caused marginalization of agricultural land, changing the diversity of European landscapes into the direction of two main types: Intensive Agriculture and Abandoned land.

ENERGY

Abandonment can be positive for nature, but this is not necessarily so. Land abandonment increases the risk of fire in the Mediterranean Region, causes a decline of small-scale landscape diversity and can also cause decrease in species diversity.

All energy types have potential impacts on the natural environment to varying degrees at all stages of use, from extraction through processing to end use. Generating energy from any source involves making the choices between impacts and how far those impacts can be tolerated at the local and global scale. This is especially of importance for nuclear power, where there are significant risks of radioactive pollution such as at Chernobyl.

Shell Oil Company and IUCN have jointly drafted environmental regulations for oil-exploitation in Arctic areas of Siberia. Other oil companies are aware of this and use these environmental regula- tions voluntarily for developing oil fields. Into the future the sustainability of the natural environment will be improved as trends away from damaging energy uses and extractive methods reduce and whilst real cost market forces and the polluter pays principle take effect.

FISHERIES

The principle of the fisheries sector is towards sustainable catches of wild aquatic fauna. The principle environmental impact associated with fisheries activities is the unsustainable har- vesting of fish stocks and shellfish and has consequences for the ecological balance of the aquatic environment. The sector is in a state of "crisis", with over capacity of the fleet, overexploitation of stocks, debt, and marketing problems.

Growing aquaculture industry may increase water pollution in western Europe, and is appearing to be a rising trend in the Mediterranean and Central/East Europe.

Fishing activities have an impact on cetaceans and there is concern that large numbers of dolphins, and even the globally endangered Monk seal, are being killed.

FORESTRY

Compared to other landuses, forest management has the longest tradition in following sustainable principles due to which over 30 per cent of Europe is still covered with trees. Without such an organised approach, forests are likely to have already disappeared from Europe's lowlands. However, as an economic sector, forestry has also impacted severely on the naturalness of Europe's forests: soils have been drained, pesticides and fertilizers applied, and exotic species planted. In many areas monocultures have replaced the original diverse forest composition. Monocultures are extremely sensitive to insect infestations, fires or wind, and so can lead to financial losses as well as biological decline. The inadequate afforestation practices characterize new trends in impacting on the sustainability of the natural environment.

INDUSTRY

Almost all forms of industry have an impact on the natural environment and its sustainability. The impact varies at different stages in the life cycle of a product, depending upon the raw materials used through to the final end use of the product for waste residue, re-use or recycling. Industrial accidents and war damage to industrial plants can also endanger the natural environment.

TOURISM AND RECREATION

Tourism and recreation impact in various ways on the natural environment. On the one hand, natural areas form the very basis of many touristic attractions by highlighting scenic value or exceptional encounters with fauna and flora. However, some forms of tourism can be extremely detrimental to ecologically sensitive areas, resulting in habitat degeneration or destruction, in the disturbance or hunting even rare or threatened species. The pressure from short holiday seasons and specific, sometimes small, locations of touristic interest result in conflicting land-uses, such as in the Alpine regions, at Mediterranean beaches and along many banks of inland waters.

TRANSPORT AND INFRASTRUCTURE

Transport is perhaps the major contributor to pollution in the world today, particularly global envi- ronmental issues such as the greenhouse effect. The key impacts of transportation include frag- mentation of habitats and species and genetic populations, disruption of migration and traffic mortalities to wildlife. Since the 1970s transport has become a major consumer of non-renewable resources, 80 per cent of oil consumption coming from road transport.

CHEMICAL POLLUTANTS

Pollutants pose a real or potential threat to wildlife on many industrial sites. Many substances have been identified as being toxic to plants and animals, and some of these are likely to occur in industrial effluents or are used in industrial processes. Major groups of toxic pollutants include various metals (including mercury, lead and cadmium), a host of organic compounds, reactive gases such as ammonia, anions such as cyanide and fluoride, acids and alkalis. Dissolved substances present a particular problem as they are readily taken up by aquatic life, often passively.

Pollutants need be present only in minute quantities to have an effect; heavy metal concentrations of only a few parts per million may inhibit root growth. Polluted waters support an impoverished fauna, although the few tolerant species can occasionally be numerous owing to reduced competition and predation in such situations.

The toxicity of many substances has often only been demonstrated in controlled laboratory experiments; in the field apparently polluted waters have often been found to support life. Where more than one pollutant is present in the water, as is often the case, they may interact, either within the water or within the organism. The effects of several pollutants together may be additive, antagonistic (where the combined effect on an organism is less than predicted by each pollutant's effect when found alone) or synergistic (where the combined effect is greater than predicted from the effects of each substance when found alone). In general it has proved difficult to predict accurately the effects of a potential pollutant in the field.

The effects of a pollutant may also be modified by environmental factors. A recent review found that pH was the factor most likely to influence toxicity (affecting 20 per cent of 410 chemicals investigated), that increasing temperature generally increased toxicity and that water hardness had little effect on organic chemicals.

The same review found that insects seemed to be the most sensitive group, followed by crustaceans, fishes and amphibians. Birds, which pick up pollutants mainly through consumption of contaminated prey, appear only rarely to die from excessive levels of pollutants (lead being an exception), and apparently absorb only limited quantities of some pollutants such as heavy metals. However, the metabolism of such substances probably puts those birds living in polluted environments under much greater physiological stress, making them much more susceptible to cold weather and disease, and potentially affecting their reproductive success.

Should water quality improve in a previously polluted wetland, invertebrate populations can recover quite rapidly, while higher plants respond much more slowly and fish populations are the last to recover. It may be that only a particular stage in the life cycle of an organism is susceptible to a pollutant.

A review of toxicity tests on fish demonstrated that the early life stages were particularly susceptible: hatchability of eggs was affected in 19 per cent of tests, survival of larvae in 57 per cent, growth of larvae in 36 per cent, reproduction in 30 per cent, adult survival in 13 per cent and adult growth in only 5 per cent. So, while individuals of a species may survive in polluted waters the species as a whole may die out if those conditions persist.

SEDIMENT AS A CHEMICAL POLLUTANT

The role of sediment in chemical pollution is tied both to the particle size of sediment, and to the amount of particulate organic carbon associated with the sediment. The chemically active fraction of sediment is usually cited as that portion which is smaller than 63 μm (silt + clay) fraction. For phosphorus and metals, particle size is of primary importance due to the large surface area of very small particles. Phosphorus and metals tend to be highly attracted to ionic exchange sites that are associated with clay particles and with the iron and manganese coatings that commonly occur on these small particles.

Many of the persistent, bioaccumulating and toxic organic contaminants, especially chlorinated compounds including many pesticides, are strongly associated with sediment and especially with the organic carbon that is transported as part of the sediment load in rivers. Measurement of phosphorus transport in North America and Europe indicate that as much as 90 per cent of the total phosphorus flux in rivers can be in association with suspended sediment.

The affinity for particulate matter by an organic chemical is described by its octanol-water partitioning coefficient (K_{OW}). This partitioning coefficient is well known for most organic chemicals and is the basis for predicting the environmental fate of organic chemicals. Chemicals with low values of K_{OW} are readily soluble, whereas those with high values of K_{OW} are described as"hydrophobic" and tend to be associated with particulates. Chlorinated compounds such as DDT and other chlorinated pesticides are very hydrophobic and are not, therefore, easily analysed in water samples due to the very low solubility of the chemical. For organic chemicals, the most important component of the sediment load appears to be the particulate organic carbon fraction which is transported as part of the sediment. Scientists have further refined the partitioning coefficient to describe the association with the organic carbon fraction (K_{OC}).

Another important variable is the concentration of sediment, especially the <63 μm fraction, in the water column. Even those chemicals that are highly hydrophobic will be found in trace levels in soluble form.

Where the suspended load is very small (say, less than 25 mg/l), the amount of water is so large relative to the amount of sediment that the bulk of the load of the chemical may be in the soluble fraction.

Unlike phosphorus and metals, the transport and fate of sediment-associated organic chemicals is complicated by microbial degradation that occurs during sediment transport in rivers and in deposited sediment.

Nevertheless, the role of sediment in the transport and fate of agricultural chemicals, both for nutrients, metals, and pesticides is well known and must be taken into account when monitoring for these chemicals, and when applying models as a means of determining optimal management strategies at the field and watershed level.For this reason, models using the"fugacity" concept (uses the partitioning characteristics of chemicals as a basis for determining the environmental compartment - air, sediment, water, biota - in which the chemical is primarily found) has proven effective in predicting the environmental pathways and fate of contaminants.

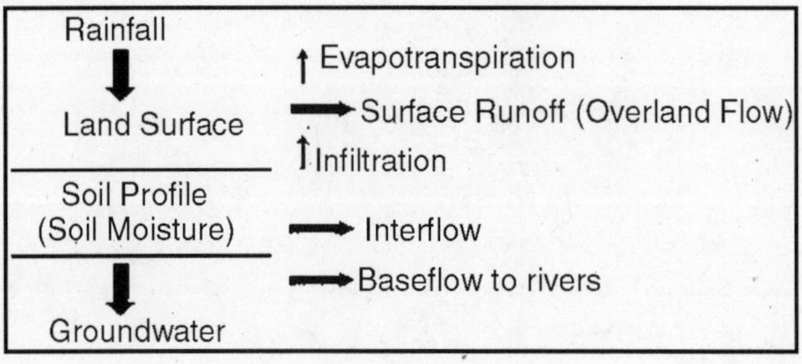

Fig. Schematic Diagram Showing the Major Processes that Link Rainfall and Run-off

- *Conclusion*: The role of sediment as a chemical pollutant is a function of the chemical load that is carried by sediments.

Organic chemicals associated with sediment enter into the food chain in a variety of ways. Sediment is directly ingested by fish however, more commonly, fine sediment (especially the carbon fraction) is the food supply for benthic (bottom dwelling) organisms which, in turn, are the food source for high organisms. Ultimately, toxic compounds bioaccumulate in fish and other top predators. In this way, pesticides that are transported off the land as part of the run-off and erosion process, accumulate in top predators including man.

KEY PROCESSES: PRECIPITATION AND RUN-OFF

The major characteristic of non-point source pollution is that the primary transfer mechanisms from land to water are driven by those hydrological processes that lead to run-off of nutrients, sediment, and pesticides.

This is important, not only to understand the nature of agricultural pollution, but also because modelling of hydrological processes is the primary mechanism by which agriculturalists estimate and predict agricultural run-off and aquatic impacts. Except where agricultural chemicals are dumped directly into watercourses, almost all other non-point source control techniques in agriculture involve control or modification of run-off processes through various land and animal (manure) management techniques.

In large parts of the world, precipitation is in the form of rain. However, in those areas where precipitation is in the form of snow, the science becomes more complex. Nevertheless, control measures, whether for areas subject to rain or snow can be easily summarized. Therefore, for the purpose of this publication, focus will be on the relationships between rainfall and run-off. While the practice of hydrology can be quite theoretical, the principal concepts are easily understood.

Rainfall

The primary controlling factor is the rate (intensity) of rainfall. This controls the amount of water available at the ground surface, and is closely related to measures of energy that are used in many mathematical formulations to calculate soil detachment by rain drops. Soil detachment makes soil particles available for sediment run-off.

Soil Permeability

Permeability is a physical characteristic of a soil and is a measure of the ability of the soil to pass water, under saturated conditions, through the natural voids that exist in the soil.

Permeability is a function of soil texture, mineral and organic composition, etc.. In contrast, "porosity" is the measure of the amount of void space in a soil; however, permeability refers to the extent to which the porosity is made up of interconnecting voids that allow water to pass through the soil. As an example, styrofoam is highly porous but impermeable, whereas a sponge is both porous and permeable.

Infiltration

Infiltration rate, the rate at which surface water passes into the soil (cm/hr), is one of the most common terms in hydrologic equations for calculating surface run-off.

Infiltration is not identical to permeability; it is mainly controlled by capillary forces in the soil which, in turn, reflect the prevailing conditions of soil moisture, soil texture, degree of surface compaction, etc. Infiltration will vary between and within rainfall events, depending upon factors such as antecedent soil moisture, nature of vegetation, etc. In general, infiltration rate begins at a high value during a precipitation event, and decreases to a small value when the soil has become saturated.

Surface Run-off

This is the amount of water available at the surface after all losses have been accounted for. Losses include evapotranspiration by plants, water that is stored in surface depressions caused by irregularity in the soil surface, and water that infiltrates into the soil.

The interaction between infiltration rate and precipitation rate mainly governs the amount of surface run-off. Intense rainstorms tend to produce much surface run-off because the rate of precipitation greatly exceeds the infiltration rate. Similarly, in areas of monsoonal rain and tropical storms, the length and intensity of precipitation frequently exceeds infiltration capacity. Destruction of protective surface vegetation and compaction of the soil, especially in tropical environments, leads to major erosional phenomena due to the amount of surface run-off. Except for nitrogen which is usually found in groundwater in agricultural areas, surface run-off is the primary contributor of agricultural chemicals, animal wastes, and sediment to river channels.

Interflow

Sometimes called "throughflow") Because soil horizons have different levels of permeability not all water in the soil will move downward into the groundwater. The residual water in the soil will move along the soil horizons, parallel to the ground surface. Interflow usually emerges near the bottom of slopes and in valley bottoms. Therefore, identification of these hydrologically active zones is an important part of agricultural non-point source control measures. Interflow is the mechanism which has also been linked to soil piping, a potentially destructive characteristic in some soils by which shallow "pipes" form naturally in the soil and are enlarged by interflow to the point where they collapse causing gullies in the agricultural surface.

Groundwater

Groundwater is supplied by water which passes through the soil horizons into the parent material and/or bedrock underlying the soil. Groundwater tends to flow towards rivers channels where it emerges and supports stream flow during periods of little or no rain.

This component of stream flow is called "base flow". The chemistry of base flow reflects the soil and bedrock geochemistry, plus any agrochemicals that have been leached into the groundwater.

Snowmelt

The phenomenon of snowmelt greatly complicates prediction of agricultural pollution using conventional hydrologic models. Snowmelt, by itself, is not normally a major producer of surface run-off.

However, the combination of spring rain and snowmelt on frozen or thawing soils can produce serious erosional problems.

Snowmelt tends to contribute greatly to agricultural non-point source pollution by carrying to adjacent streams the animal wastes, sludges, and other wastes that were spread on frozen agricultural soils during the winter period. Correct management of animal wastes in regions of frozen ground has major beneficial effects on water quality.

CHEMICAL POLLUTION

According to the Environmental Protection Agency, people are exposed to a variety of chemicals on a daily basis. Chemicals are used in a variety of locations visited on a regular basis including schools, grocery stores, and dry cleaners. Chemicals are also common ingredients in household cleaners and can be found in a number of consumer products.

The EPA collects data on chemicals that are used and actively works with industries to minimize the diminish the damage caused to the environment. When chemicals are released into the environment, pollution occurs. This can occur when chemicals are released, either knowingly or unknowingly, into water, air, or land.

AGRICULTURAL CHEMICALS

Agriculture is a large source of chemical pollution. There are a variety of chemicals and pesticides that are used in agricultural production. These chemicals can seep into the soil and eventually make their way into groundwater and our food supply. According to the Civil Engineering Department at Virginia Tech phosphates, herbicides, nitrates, bacteria, and pesticides are common pollutants that are introduced to the environment through agricultural means.

WATER POLLUTION

Water pollution can be caused through a variety of means. Water pollution is often tied to agricultural pollution due to the large amount of pesticides and other chemicals that often enter water supplies. Runoff from livestock feedlots and pastures also contributes to this type of pollution. Another major source of water pollution is oil spills and emissions from aquatic vehicles such as boats and jet skis. According to the University of Michigan, this water pollution can be very harmful to aquatic life. Aquatic plants and fish can suffer from respiratory problems and other issues. Fish is a major source of the American diet and chemical pollution can severely endanger the fish population. In some instances it could also be possible for chemicals to be transferred to the general population via fish consumption.

AIR POLLUTION

Air pollution is possibly the most commonly thought of form of chemical pollution. An article published by National Geographic discusses the various pathways through which air pollution can occur. Chemical air pollution that is produced by manufacturing businesses and other industries is the most noted form of air pollution.

Cars and planes also give off emissions that can pollute the air. Carbon dioxide is given off as these forms of transportation burn fossil fuels. Although plants and other living things give off carbon dioxide, the levels given off by modes of transportation is significantly greater and has a greater effect on the atmosphere. The National Geographic article also notes that volcanic eruptions and gases emitted from swamps and livestock also contribute minimal amounts to air pollution.

Cleanup

Cleanup of environmental pollution can be time consuming, intricate, and expensive. The tactics chosen, time that needs to be invested in the process, and the techniques employed depend on the type of chemical in question and the size of the area affected. According to an article published on a website managed by Advameg, there are several government agencies that aid the cleanup of environmental accidents. Many of these cleanup efforts are funded by taxes and fees paid by residents. Volunteer efforts are also beneficial in some instances, although chemical cleanups often require professional assistance and specific safety protocol.

Prevention

Prevention is the best measure for controlling chemical pollution. The EPA actively works with industries to help monitor gas emissions and disposal of hazardous chemicals.

There are measures taken to ensure that many of the chemicals and gases used in everyday processes are disposed of and recycled correctly.

Individuals are encouraged to reduce their "carbon footprint", essentially the amount of pollution they contribute to the environment, by recycling, choosing environmentally-friendly modes of transportation such as walking and biking, and living an overall green and environmentally conscious lifestyle.

SHOULD WILDLIFE BE ATTRACTED TO CONTAMINATED WETLANDS

Before all else, consideration needs to be given to whether the water available to create a wetland is of an appropriate quality to warrant encouraging wildlife. Unfortunately, our understanding of the way pollutants affect organisms and interact with each other is far from comprehensive, making it difficult to predict the problems that an industrial wetland might pose to wildlife.

In trying to assess whether wildlife will be at risk at a particular site, consideration must be given to a variety of factors:

- The types of wildlife that might use a site;
- The chemistry and characteristics of the water;
- The characteristics of any potentially toxic substances in the wetland - toxicity, persistence, bio-availability, and the potential for bio-accumulation.

Different groups of plants and animals vary greatly in the way they respond to the presence of contaminants. Organisms living in the water column, such as aquatic plants, fish and invertebrates, are likely to be significantly affected by water chemistry. These organisms often take up pollutants dissolved in the water, plants occasionally accumulating Very high concentrations, and are vulnerable to oxygen depletion resulting from high BOD.

Submerged plants are also adversely affected by reduced light penetration caused by suspended solids; Fennel Pondweed seems to be the most tolerant species in this respect. By contrast, emergent plants absorb most substances, including pollutants, from the sediment in which they are rooted. Reedmace can apparently become acclimatised to sediments contaminated with metals, despite metal concentrations sometimes accumulating in the root system (very little is stored in the above-ground portions of the plant). Many metal compounds probably pass straight through the guts of birds undigested, while most other metals can apparently be excreted by birds. Complex organic substances, such as organochlorines, certainly can cause chronic conditions in birds, while organo-lead compounds are lethal in even modest concentrations. It is, therefore, difficult to generalise as to whether a water source will be harmful to wildlife; in some wetlands aquatic plants and many invertebrates may be excluded owing to the pollution levels while birds apparently prosper.

Probably the factor that should be given most attention is the potential bio-availability of known toxic substances. Inert, immobile forms of such substances may not present a problem to wildlife. Metals such as lead and copper, for example, will remain as insoluble and immobile sulphides as long as they are trapped within anoxic sediments. The pH will be critical in this respect as it affects the mobility and bio-availability of most metals, many of which can go into solution under acid conditions. The activities of benthic invertebrates may lead to the disturbance or chemical transformation of contaminated sediments.

Consequently, it is possible for the bio-availability of a substance to change with time, where, for instance, the water chemistry is affected by a fluctuating management regime. A possible example of this effect was seen in the Mersey Estuary where there was a series of bird kills between 1979 and 1981 thought to have been caused by discharges of alkyl lead compounds. These discharges had been on-going for many years, suggesting that some other factor must have triggered the sudden lethal effects. One possibility is that a reduction in general pollution levels within the estuary resulted in increased invertebrate populations in the discharge area which in turn increased the birds' consumption of prey contaminated with the lead compounds.

The likelihood of a substance passing up a food-chain also varies tremendously. Sampling undertaken at Werribee wastewater treatment works in Australia demonstrated that although metal contamination in the wastewater was taken up by the plants growing in the grass plots, very little of this accumulated in the tissues of the livestock that grazed them, and that copper levels were actually depressed in these stock. Ingeneral, it is the lipid-soluble toxins, such as the methylated forms of mercury and lead and organochlorines, that show a tendency to accumulate (bio-magnify) as they pass up a food-chain, and hence pose most concern with regard to the welfare of vertebrates.

Polluted wetlands obviously are not universally good for wildlife, as evidenced by the impoverished species diversity that is often associated with them. That is not to say, however, that contaminated wetlands cannot provide valuable habitat for some forms of wildlife, as demonstrated by Features 12.1 Some effluents will contain such high levels of toxic substances that common sense will dictate that they are dealt with sensitively and in such a way that wildlife should not come into contact with them. A more difficult ethical issue is whether wildlife should be deterred or excluded from sites where they might be exposed to substances that can cause chronic rather than lethal conditions. At the present time there are no hard and fast rules and each situation should be considered individually.

Until such time as our knowledge suggests otherwise, it is recommended that wildlife is not attracted, or is even deterred from, situations where:

- The water contains substances in concentrations that are likely to be lethal to birds and mammals;
- Sediments are likely rapidly to accumulate toxic substances in forms that are, or are likely to become, soluble or otherwise mobile and therefore available for biological up-take. Sites where accumulation of such substances is likely to occur only at a slow rate, should be monitored to ensure that critical levels are not reached. If such levels are reached then the contaminated sediments should be removed or the site design or management modified to deter these groups of wildlife;
- Known toxic substances, present in significant concentrations, are likely to bio-magnify as they are passed up a food-chain.

POLLUTION THROUGH AGRICULTURAL WASTES

Wastes and residues resulting from diverse agricultural practices for raising the crops and animals are collectively called agricultural wastes. In other words, refuse of any form from agricultural operations of any kind is an agricultural waste. In India, main sources of wastes are straw and bran of rice and wheat, maize cobs, leftovers from pulses etc. The generation of paddy husk is around 18 million tonnes annually and fruit and vegetable wastes are estimated to be 25,000 tonnes approximately. The manure production from our animal population is 906 million tonnes per annum. Accumulation of these wastes and the inability of these wastes to disintegrate quickly is causing a lot of pollution problem. The most common type of agricultural wastes are briefly described here. For more details about various types of agriculture and farm wastes consult booklets No. 31 on "Organic Manures" No. 63 on "Farm Yard Manure" and No. 66 on "Green manuring". Any source of manure when not used becomes a pollutant.

LIQUID WASTES

Liquid wastes from the feedlot industry seriously contaminates the water table. Ammonia of these feedlot wastes being readily soluble in water, they return to nearby lakes and streams. The feedlot runoffs adds disease causing orgimisms to water. So, careful management of liquid wastes is essential they contain celluloses fibres and other undigested foods which can be safely fed to the animals.

ANIMAL WASTES

Animal dung availability in India is significantly high going up to 906 million tonnes per year which includes wet cattle dung (860 million tonnes), sheep and goat wastes (95 million tonnes) and poultry wastes (one million tonne). These animal wastes are major pollutants once they enter into water supplies. When confinement areas are cleaned or when these wastes are carried away by the runoff water contaminating the water ways environmental pollution and health hazards occur. The availability of commercial fertilizers rich in nutrients and easy to handle, has forced the farmers to neglect these natural wastes instead of reusing them. The most efficient method of avoiding pollution by animal wastes is the use the biologically active systems of aerobic and anaerobic digestion. Anaerobic digestion of wastes leads to production of methane gas which can be used as energy.

Anaerobic digestion can be combined or followed by aerobic oxidation, to give the effluent an adequate quality for discharging it into rivers or main drainage systems. Whatever may be the method, the manure should be returned to the soil with enriched nutrient content. The wastes can be better recycled by suitable methods like composting, dehydration, incineration, etc. In a certain sense the application of waste while they are fresh is better than application of rotten matter in order to avoid the loss of nutrients by leacing, volatalisation, solubilisation etc. But, practically, application of fresh manure is difficult and need a proper follow up care after addition. After addition the process of decomposting starts by which heat is generated and in lack of moisture the heat increases soil temperature which is harmful to beneficial organisms.

HARVEST WASTES

These include the straw left over from harvesting of cereals, millets, pulses, oil seeds and vegetables. These wastes acts as carrier of the diseases unless they are managed carefully. They can be used as a bedding material and also as an absorbant for liquid wastes and also in cleaning the oil spills from surfaces and shore areas. These wastes can be used as cattle feed or as mulch or ploughed into the soil itself which eventually decays and provides nutrients to the soil.

Plant Residues

Plant residues from crops and orchards create environmental pollution when they harbour plant diseases or pests. All these release carbon, oxygen, nitrogen, oxides and hydrocarbons when burnt, leading to air pollution. Burning can release around 8.3 per cent carbon-monoxide 5.3 per cent, hydrocarbons and 1.5 per cent nitrogen oxide. Instead of burning plant debris can be best used as a mulch which also reduces erosion.

Food Processing Wastes

Food processing wastes are as varied as the goods processed. Pea pods, tomato wastes and potato peels, soybean hulls, peanut shells and cotton wastes, oil seed wastes after extracting oil, sugar cane trasnafter extracting the juice, water from cheese making and dairy plants, any chemical used in food processing etc. are some of the common pollutants from agricultural sector. Water pollution is the most serious problem that is created by these pollutants. When they are dumped into surface water sources they reduce the oxygen content in water.

So these products can be treated anaerobically or aerobically before they are used as manure or animal feeds. These wastes contain considerable amounts of plant and animal nutrients. Through the anaerobic digestion we can also generate considerable amount of energy for our daily usages. The chemical composition of these wastes greately varies and the processing methods too vary greatly but by suitable categorization and processing, the resource value can be increased as well as the problem of pollution can be solved.

POLLUTION THROUGH FERTILIZERS

The fertilizers, no doubt, are important in intensive agriculture. When applied to soil, they meet the nutrient needs of the crop. The fertilizer use efficiency depends on the type of management practices adopted. Not all the quantity of applied fertilizer is taken up by the crop but some quantity is retained by soil, which can be washed down by erosion or by drainage water into water sources or can also be lost into atmosphere as a gas. All these up to a certain Ievel are safe and if they exceed beyond certain limit they pollute the atmosphere, water sources and soil. The major nutrients that are supplied through fertilizers are nitrogen, phosphorus and potassium and the fate of these nutrients is briefly described here.

Nitrogen

Nitrogen is mostly taken up by crops either in the form of ammonia or nitrate. Nitrate form is taken up by most of the crops except paddy which takes nitrogen in the form of ammonia. Nitrate itself is not toxic. In fact it has its own medicinal importance in treating kidney stones. Nitrate becomes a problem only when it is converted into nitrite. When applied to soil as a fertilizer, being soluble in water some quantity will be lost in runoff water leading to pollution of surface water sources like lakes, ponds, tanks etc. In India, farmers have a practice of letting off water from one rice field to another field during flood irrigation. This leads to the loss of applied nitrogen.

Nitrogen is also lost by volatalisation, leaching, denitrification and fixation. For more details about these process consult booklets on "Fertilizers: An Introduction" (No.5), "Fertilizer Recommendations" (No. 19), "Fertilizer Application" (No. 23), Fertilizer Management in Rice (No. 36), "Fertilizer Use in Dry Lands (No.. 43) and "Fertilizer Consumption and Extension" (No. 51). All these types of losses are leading to pollution which can be controlled by various agronomic and management practices.

The most important among all the control measures is to increase the fertilizer-use efficiency and also ensure greater utilisation of the applied as well as residual nitrogen so that less and less of it is prone to different losses that are mentioned above. Selection of suitable form of fertilizer for a particular crop under a given soil and climatic condition, application of suitable dose at right time, by a proper method in a balanced proportion are the effective agronomic practices to increase nitrogen use efficiency and decrease the effect of residual nitrogen. For more details, refer booklet on "Fertilizer Application" (No.23).

Recently popularised slow-release nitrogen fertilizers also inhibit the losses to a considerable extent. Well managed and fertilized land under efficient crop rotation will have lesser losses of nitrogen when compared to fallow land or poorly managed and. Residual nitrite causes problems to human health. A problem known as 'blue-baby syndrome' or 'methaemoglobinaemia' occurs in children. They are susceptible because their stomach do not have enough acid to inhibit the microbes that convert nitrate to nitrite. This condition proves to be fatal.

Phosphorus

Phosphate losses may take place basically through soil erosion by water and it moves down the profile. Since the soil solution is generally low in phosphorus, its downward movement is also insignificant. Still the loss of this nutrient by soil erosion can be controlled by different crop rotations.

Potassium

Mostly nutrient loss of potassium occurs by soil erosion and leaching.

control measures like proper crop rotation, avoiding fallows, soil and water conservation measures easily reduce the potassium losses and the pollution effect created by it. Hence, many fertilizers, which are found to be beneficial, for increasing yield have also the harmful effect by some sort of residue. As already given, when ammonium sulphate is applied, sulphate may be left as a residue so do with sodium nitrate where sodium will be left as a residue. These residues increase year after year and pollute the soil and the environment.

POPULATION GROWTH AND ENVIRONMENTAL QUALITY

There is a long history of study and debate about the interactions between population growth and the environment. According to the British thinker Malthus, for example, a growing population exerts pressure on agricultural land, causing environmental degradation, and forcing the cultivation of land of poorer and poorer quality.

This environmental degradation ultimately reduces agricultural yields and food availability, causes famines and diseases and death, thereby reducing the rate of population growth. Population growth, because it can place increased pressure on the assimilative capacity of the environment, is also seen as a major cause of air, water, and solid-waste pollution. The result, Malthus theorized, is an equilibrium population that enjoys low levels of both income and environmental quality. Malthus suggested positive and preventative forced control of human population, along with abolition of poor laws.

Malthus theory, published between 1798 and 1826, has been analyzed and criticized ever since. The American thinker Henry George, for example, observed with his characteristic piquancy in dismissing Malthus: "Both the jayhawk and the man eat chickens; but the more jayhawks, the fewer chickens, while the more men, the more chickens." Similarly, the American economist Julian Lincoln Simoncriticised Malthus's theory. He noted that the facts of human history have proven the predictions of Malthus and of the Neo-Malthusians to be flawed. Massive geometric population growth in the 20th century did not result in a Malthusian catastrophe.

The possible reasons include: increase in human knowledge, rapid increases in productivity, innovation and application of knowledge, general improvements in farming methods (industrial agriculture), mechanization of work (tractors), the introduction of high-yield varieties of wheat and other plants (Green Revolution), the use of pesticides to control crop pests.

More recent scholarly articles concede that while there is no question that population growth may contribute to environmental degradation, its effects can be modified by economic growth and modern technology. Research in environmental economics has uncovered a relationship between environmental quality, measured by ambient concentrations of air pollutants and per capita income. This so-called environmental Kuznets curve shows environmental quality worsening up until about $5,000 of per capita income on purchasing parity basis, and improving thereafter. The key requirement, for this to be true, is continued adoption of technology and scientific management of resources, continued increases in productivity in every economic sector, entrepreneurial innovation and economic expansion.

Other data suggests that population density has little correlation to environmental quality and human quality of life. India's population density, in 2011, was about 368 human beings per square kilometer. Many countries with population density similar or higher than India enjoy environmental quality as well as human quality of life far superior than India.

For example: Singapore (7148 /km^2), Hong Kong-China (6349 /km^2), South Korea (487 /km^2), Netherlands (403 /km^2), Belgium (355 / km^2), Japan (337/ km^2), England (395 /km^2), the state of Florida (353 / km^2), the state of New York (412 / km^2), the state of Massachusetts (840 / km^2), the state of North Rhine-Westphalia (523 / km^2), the region of Île-de-France (974 / km^2), and the region of Lombardy (417 / km^2).

WATER POLLUTION

Methods of controlling water pollution fall into three general categories: physical, chemical, and biological. For example, one form of water pollution consists of suspended solids such as fine dirt and dead organisms. These materials can be removed from water by simply allowing the water to sit quietly for a period of time, thereby allowing the pollutants to settle out, or by passing the water through a filter. (The solid pollutants are then trapped in the filter.)

Chemical reactions can be used to remove pollutants from water. For example, the addition of alum (potassium aluminum sulfate) and lime (calcium hydroxide) to water results in the formation of a thick, sticky precipitate. When the precipitate begins to settle out, it traps and carries with it solid particles, dead bacteria, and other components of polluted water. Biological agents can also be used to remove pollutants from water. Aerobic bacteria (those that need oxygen to survive) and anaerobic bacteria (those that do not require oxygen) attack certain chemicals in polluted water and convert them to a harmless form.

SOLID POLLUTANTS

Solid pollutants consist of garbage, sewage sludge, paper, plastics, and many other forms of waste materials. One method of dealing with solid pollutants is simply to bury them in dumps or landfills. Another approach is to compost them, a process in which microorganisms turn certain types of pollutants into useful fertilizers. Finally, solid pollutants can also be incinerated (burned).

Trade and Environment

The main goal is to make the complex relationship between the environment and international trade more understandable from the developing countries' perspective. It also aims to dispel the idea that the relationship between trade and environment can easily be described as either negative or positive. It is extremely complex and varies from country to country. There are both threats and opportunities in this relationship for developing countries pursuing economic development and environmental protection.

The conclusions that can be drawn from this that from the developing countries' perspective is to exploit the opportunities and reduce the threats, and in so doing, to maximize the net positive contribution that trade can make sustainable development. A broader and clearer understanding of the linkages between trade, environment, and development is a prerequisite for seizing those opportunities and reducing those threats.

On the other hand, environmental requirements however, may be used for protectionist purposes and may operate as a non-tariff barrier. In future, there is a likelihood of more trade embargos in the name of environment, especially in cases where countries fail to comply with certain environmental standard requirements. Thus, the outcome of the various case studies has the potential for strengthening the hand of protectionist lobbies, in the name of conservation and protection of the environment.

Finally, the threat of embargo can also be used to make countries adhere to certain environmental policies and regulations, which they would otherwise not adopt due to various socio-economic compulsions. Moreover, these environmental friendly policies demand the attention and cooperation of international institutions such as the World Bank as well as NGOs and other grassroots environmental groups. Protecting the environment is a luxury good, often only afforded to the wealthy countries.

Therefore, these policies should be coupled with ones that encourage the transfer and exchange of environmentally friendly technologies and the build up of industries oriented to sustainable development.

Water Pollution and Sewage

A 2007 study finds that discharge of untreated sewage is single most important cause for pollution of surface and ground water in India. There is a large gap between generation and treatment of domestic wastewater in India. The problem is not only that India lacks sufficient treatment capacity but also that the sewage treatment plants that exist do not operate and are not maintained. Majority of the government-owned sewage treatment plants remain closed most of the time due to improper design or poor maintenance or lack of reliable electricity supply to operate the plants, together with absentee employees and poor management. The wastewater generated in these areas normally percolates in the soil or evaporates. The uncollected wastes accumulate in the urban areas cause unhygienic conditions and release pollutants that leaches to surface and groundwater.

A 1992 World Health Organization study is claimed to have reported that out of India's 3,119 towns and cities, just 209 have partial sewage treatment facilities, and only 8 have full wastewater treatment facilities. Downstream, the untreated water is used for drinking, bathing, and washing. A 1995 report claimed 114 Indian cities were dumping untreatedsewage and partially cremated bodies directly into the Ganges River. Open defecation is widespread even in urban areas of India. This situation is typical of India as well as other developing countries. According to another 2005 report, sewage discharged from cities and towns is the predominant cause of water pollution in India. Investment is needed to bridge the gap between 29000 million litre per day of sewage India generates, and a treatment capacity of mere 6000 million litre per day. A large number of Indian rivers are severely polluted as a result of discharge of domestic sewage.

The Central Pollution Control Board, a Ministry of Environment & Forests Government of India entity, has established a National Water Quality Monitoring Network comprising 1429 monitoring stations in 27 states and 6 in Union Territories on various rivers and water bodies across the country. This effort monitors water quality year round. The monitoring network covers 293 rivers, 94 lakes, 9 tanks, 41 ponds, 8 creeks, 23 canals, 18 drains and 411 wells distributed across India. Water samples are routinely analyzed for 28 parameters including dissolved oxygen, bacteriological and other internationally established parameters for water quality. Additionally 9 trace metals parameters and 28 pesticide residues are analyzed. Biomonitoring is also carried out on specific locations.

The scientific analysis of water samples from 1995 to 2008 indicates that the organic and bacterial contamination are severe in water bodies of India. This is mainly due to discharge of domestic wastewater in untreated form, mostly from the urban centres of India.

In 2008, the water quality monitoring found almost all rivers with high levels of BOD. The worst pollution, in decreasing order, were found in river Markanda (590 mg O/l), followed by river Kali (364), river Amlakhadi (353), Yamuna canal (247), river Yamuna at Delhi (70) and river Betwa (58). For context, a water sample with a 5 day BOD between 1 and 2 mg O/L indicates a very clean water, 3 to 8 mg O/L indicates a moderately clean water, 8 to 20 indicates borderline water, and greater than 20 mg O/L indicates ecologically-unsafe polluted water.

The Mithi River, which flows through the city of Mumbai, is heavily polluted. The levels of BOD are severe near the cities and major towns. In rural parts of India, the river BOD levels were sufficient to support aquatic life.

Total coliform and fecal coliform densities in the rivers of India range between 500 to 100,000 MPN/100 ml. The presence of coliform suggests that the water is being contaminated with the fecal material of humans, livestocks, pets and other animals. Rivers Yamuna,Ganga, Gomti, Ghaggar, Chambal, Mahi, Vardha are amongst the other most coliform polluted water bodies in India. For context, coliform must be below 104 MPN/100 ml, preferably absent from water for it to be considered safe for general human use, and for irrigation where coliform may cause disease outbreak from contaminated-water in agriculture.

In 2006, 47 percent of water quality monitoring stations in India reported a total average annual coliform levels above 500 MPN/100 ml. During 2008, 33 percent of all water quality monitoring stations reported a total coliform levels exceeding those levels, suggesting recent effort to add pollution control infrastructure and upgrade treatment plants in India, may be reversing the water pollution trend. Treatment of domestic sewage and subsequent utilization of treated sewage for irrigation can prevent pollution of water bodies, reduce the demand for fresh water in irrigation sector and become a resource for irrigation. Since 2005, Indian wastewater treatment plant market has been growing annually at the rate of 10 to 12 percent. The United States is the largest supplier of treatment equipment and supplies to India, with 40 percent market share of new installation. At this rate of expansion, and assuming the government of India continues on its path of reform, major investments in sewage treatment plants and electricity infrastructure development, India will nearly triple its water treatment capacity by 2015, and treatment capacity supply will match India's daily sewage water treatment requirements by about 2020.

Water resources have not been linked to either domestic or international violent conflict as was previously anticipated by some observers. Possible exceptions, notes a 2004 report, include some communal violence related to distribution of water from the Kaveri River and political tensions surrounding actual and potential population displacements by dam projects, particularly on the Narmada River. A 1997 article claimed Punjab is another hotbed of pollution, for example, Buddha Nullah, a rivulet which run through Malwaregion of Punjab, India, and after passing through highly populated Ludhiana district, before draining into Sutlej River, a tributary of the Indus river, is today an important case point in the recent studies, which suggest this as another Bhopal in making.

A joint study by PGIMER and Punjab Pollution Control Board in 2008, revealed that in villages along the Nullah, calcium, magnesium, fluoride, mercury, beta-endosulphan and heptachlor pesticide were more than permissible limit (MPL) in ground and tap waters. Plus the water had high concentration of COD and BOD (chemical and biochemical oxygen demand), ammonia, phosphate, chloride, chromium, arsenic and chlorpyrifos pesticide.

The ground water also contains nickel and selenium, while the tap water has high concentration of lead, nickel and cadmium. The Hindon River, which flows through the city of Ghaziabad, highly polluted and groundwater of this city has coloured and poisoned by industrial effluents, Hindon Vahini is strongly opposing of water pollution activities. Flooding during monsoons worsens India's water pollution problem, as it washes and moves all sorts of solid garbage and contaminated soils into its rivers and wetlands.

The annual average precipitation in India is about 4000 billion cubic metres. From this, with the state of Indian infrastructure in 2005, the available water resource through the rivers is about 1869 billion cubic meters. Accounting to uneven distribution of rain over the country each year, water resources available for utilization, including ground water, is claimed to be about 1122 billion cubic meters. Much of this water is unsafe, because pollution degrades water quality. Water pollution severely limits the amount of water available to Indian consumer, its industry and its agriculture.

The Ganges

To know why 1,000 Indian children die of diarrhoeal sickness every day, take a wary stroll along the Ganges in Varanasi. As it enters the city, Hinduism's sacred river contains 60,000 faecal coliform bacteria per 100 millilitres, 120 times more than is considered safe for bathing. Four miles downstream, with inputs from 24 gushing sewers and 60,000 pilgrim-bathers, the concentration is 3,000 times over the safety limit. In places, the Ganges becomes black and septic. Corpses, of semi-cremated adults or enshrouded babies, drift slowly by. — *The Economist on December 11, 2008*

More than 400 million people live along the Ganges River. An estimated 2,000,000 persons ritually bathe daily in the river, which is considered holy by Hindus. Ganges river pollution from sewage and semi-cremated remains is thus a major health risk.

The Yamuna

NewsWeek describes Delhi's sacred Yamuna River as "a putrid ribbon of black sludge" where fecal bacteria is 10,000 over safety limits despite a 15-year program to address the problem. Cholera epidemics are not unknown. NewsWeek observes India's messy democracy is particularly ill equipped to handle the conflicting pressures of rapid growth and poverty.

Even though India revised its national water policy in 2002 to encourage community participation and decentralize water management, the country's Byzantine bureaucracy ensures that it remains a "mere statement of intent." Responsibility for managing water issues is fragmented among a dozen different ministries and departments without any coordination. The government bureaucracy and state-run project department has failed to solve the problem, despite having spent many years and $500 million on this project.

GLOBAL ENVIRONMENTAL PROBLEMS

Ethics is understood to be the domain of enquiry that explores what is right or wrong, obligatory or non-obligatory, or when responsibility attaches to human Behaviour. Are there features of global environmental problems that call for classifying them as essentially ethical problems with even greater force than some local or regional environmental problems? If so, what are these features?

Why is this important? If some global environmental problems are essentially ethical problems, then some of the excuses that nations often use to justify not reducing their contributions to the global environmental problems are ethically problematic. This is so because ethical obligations entail duties and responsibilities to others with the result that national policies may not be justified on national interests alone. That is, if there are ethical obligations to others to cease causing harm, then policy options must be responsive to these obligations.

The features of global environmental problems that strongly call for classification as ethical problems. Some global environmental problems are caused by people in one part of the world but are most harshly experienced by others who are separated from those causing the problem by great space and time.

In addition, those most vulnerable to global environ-mental problems are often least responsible for causing them. Climate change is a strong example of this feature of global environmental problems. Those most vulnerable to climate include many of the poorest people in the world who are among those least responsible for emitting the greenhouse gases that are causing the problem.

The same separation between victim and cause is also true of other global environmental problems, including upper atmospheric ozone depletion, some ocean degradation problems including destruction of global fisheries, some biodiversity and forest problems, and exposure to toxic substances that are deposited around the world by long-range air transport.

In the case of global fisheries depletion, for instance, fishing trawlers from some countries often go great distances to harvest fish for domestic consumption, yet the harshest consequences of depleted global fisheries are often experienced by subsistence fisherman in poor countries far away from the home port of the fishing boats.

This separation of space and time and between cause and impact of course can also happen in more local environmental problems such as water pollution, but the distances between those causing the problem and those who suffer the consequences are most problematic at the global scale because of the magnitude of the distances at this scale.. With the rise in truly global environmental problems in the last few decades, it is now abundantly clear that Behaviour of people in one part of the world can gravely affect the human health, economic viability and overall welfare of others far away, but their suffering is almost invisible to those who cause the problem.

Several implications for ethical reasoning follow directly from this separation of distance and time between cause and effect. First, our intuitive sense of ethical responsibility for harms that we cause others is harder to trigger when those who will be harmed by our action are people we will not ever know and are from a different culture. Yet ethics and morality require us to refrain from harming all others who may suffer because of our Behaviour even if they are separated from us in time and space.

Secondly, because the actual harm that we are causing to others may be in a particular part of the world that is uniquely vulnerable to global environmental change but is unknown to us, we not only don't know the people we are harming but have no idea where the locus of the worst harm is located. For instance, particularly vulnerable villages to climate change impacts may turn on the geographical features where the village is located such as dependence on glacier water flows in the summer coupled with vulnerability to monsoon irregularities. Yet few of us know where these places are. And so in the case of global environmental problems we may be causing great harm to others who we not only do not know personally, but about whom we neither have an idea of their location nor why they are particularly vulnerable to actions we are taking. For this reason, global scale environmental problems create a huge challenge to trigger our sense of ethical responsibility, yet ethics requires us to not harm others through our actions none the less.

A related reason why the global environmental problems demand classification as ethical problems stems from the fact that some actions that cause some environmental damage are often justified on the basis of cost-benefit analysis. Yet this justification is ethically problematic when those who benefit from an action that causes harm have no relation with those who are harmed. For instance, government may tolerate inevitable pollution from a land fill by identifying the benefits that citizens receive by having a place to take their garbage. But in the case of global environmental problems, those who are harmed by actions do not enjoy the benefits that come from taking the action. A harms-benefits justification for environmental degradation, if ethically justifiable at all, only works where those who will suffer the harms and those who benefit from the action that causes the harm are in the same community of interests. In the case of global environmental problems, those who are harmed are rarely made better by the benefits of the action that cause the harm.

The second reason why global environmental problems call for classification as ethical problems stems from the fact that their consequences are often catastrophic to those who are most harmed by them. Climate change, for instance, directly threatens human life and health and resources to sustain life as well as species of plants and animals and ecosystems around the world. The harms include deaths from disease, droughts, floods, heat and intense storms and damage to homes and villages from rising oceans and intense storms, adverse impacts on agriculture, social disputes caused by diminishing natural resources, sickness from a variety of diseases, the inability to rely upon traditional sources of food, the destruction of water supplies, and the inability to live where one has lived to sustain life. In addition, the very existence of some small island nations is threatened by climate change caused seal level rise. Clearly these impacts are catastrophic for some.

Loss of global fisheries can be devastating to substance fishermen, global desertification and drought are likely to cause mass starvation, loss of atmospheric ozone can cause skin cancer in millions, and long-range transport of persistent organic chemicals can cause deadly diseases and birth defects. Because ethics requires people to be particularly careful to not to harm others when the harm is great, global environmental problems need to be seen as ethical issues.

Governments are expected to protect their citizens from life-threatening dangers. At the local, regional or national scale, citizens can petition their governments to protect them from environmental harms. But at the global level, no government exists whose jurisdiction matches the scale of global environmental problems. And so, although existing national, regional and local governments have jurisdiction over activities within their boundaries coupled to responsibilities to their citizens, they have no responsibility to those outside their boarders in the absence of binding international law. For this reason, ethical appeals to get governments to require that their citizens not harm non-citizens in other nations are necessary to prompt action.

In fact, according to a leading foreign policy theory called "realism," national governments should limit the scope of their international engagements to the pursuit of national interest alone. According to some "realist" proponents, governments should refrain from acting at the international level for ethical reasons that do not coincide with national interests. Yet global environmental problems require people in one part of the world to consider the interests of others outside the jurisdiction of their national governments. Because there is no duty of national governments to protect foreigners from global environmental problems unless they bind themselves voluntarily in treaties, an appeal to ethical and moral responsibility is particularly important to motivate national actions in regard to global scale environmental problems. Moreover, if global environmental problems actually create ethical obligations, they have ethical duties to mitigate global environmental problems

CFCS IN ATMOSPHERE

There has been a movement since the late 1970s to ban CFCs because of their destructive effect on the ozone layer. This damage was discovered by Sherry Rowland and Mario Molina, who first published a paper suggesting the connection in 1974. It turns out that one of CFCs' most attractive features-their unreactivity-has been instrumental in making them one of the most significant pollutants. CFCs' lack of reactivity gives them a lifespan which can exceed 100 years in some cases. This gives them time to diffuse into the upper stratosphere. Here, the sun's ultraviolet radiation is strong enough to break off the chlorine atom, which on its own is a highly reactive free radical. This catalyses the break up of ozone into oxygen:

$$Cl + O_3 \rightarrow ClO + O_2$$

$$ClO + O \rightarrow Cl + O_2$$

CFCs are a problem because the chlorine is regenerated at the end of these reactions, making it able to repeat this reaction millions of times.

This disrupts the ionic structure of oxygen atoms, causing their lowest ionisation energy to drop. This reaction is believed to be the cause of the ozone hole observed over the poles and upper latitudes of the Earth. Decreases in stratospheric ozone have been linked to increases in skin cancer. In 1975, Oregon enacted the world's first ban of CFCs.

By 1985, scientists observed a dramatic seasonal depletion of the ozone layer over Antarctica, which led the United Nations to take action in 1987. International attention to CFCs resulted in a meeting of world diplomats in Montreal in 1987. They forged a treaty, the Montreal Protocol, which called for drastic reductions in the production of CFCs. On March 2, 1989, 12 European Community nations agreed to ban the production of all CFCs by the end of the century. In 1990, diplomats met in London and voted to significantly strengthen the Montreal Protocol by calling for a complete elimination of CFCs by the year 2000. By the year 2010 CFCs should be completely eliminated from developing countries as well

FREON

Freon is a trade name for a group of chlorofluorocarbons used primarily as a refrigerant. The word Freon is a registered trademark belonging to Dupont. Prior to their implication as a major contributor to ozone depletion (and a minor contributor to the greenhouse effect), freons were also used as aerosol propellants and to foam up polymer materials.

Freon was developed as a safer alternative to toxic gases such as, ammonia (NH_3), methyl chloride (CH_3Cl), and sulfur dioxide (SO_2), that were used as refrigerants prior to the discovery of freon. Despite the controversy, R12 works well and is very safe. Its inventor, Thomas Midgley, actually breathed in a lungful and exhaled over a candle to demonstrate its nontoxicity and nonflammability. Before CFCs were banned by developed nations, a 12 ounce (340 g) can of freon (R12) cost about $1; following the ban the cost of freon rose dramatically. People now go through considerable trouble to recycle freon from existing sources. As of 2002, freon cost about $85 per lb ($187 per kg). Freon is most frequently sold in 12 ounce (340 g) cans and 30 pound (13.6 kg) canisters.

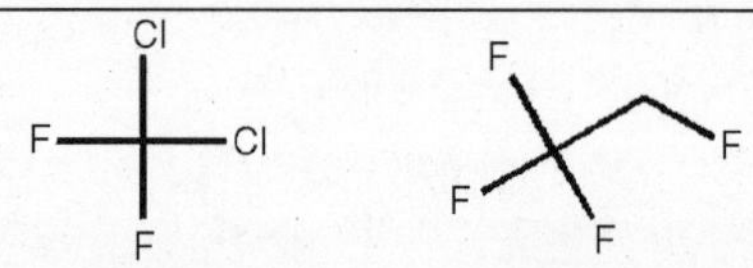

Dichlordifuoromethane "Freon 12"   1,1,1,2-Tetrafluoroethane "Suva 134a"

Freon 12 ® (also known as R–12) was a popular re-frigerant before it was banned. The chemical formula is $CCl_2F_2$and its proper chemical name is dichloro-difluoromethane. When people refer to freon without specifying a number, it is likely that they mean freon-12.

Suva 134a ® (also known as R–134 or R–134a) is the most common replacement for R–12. Unlike R–12, which is a chlorofluorocarbon, R–134 is a hydro fluorocarbon. The chemical formula is CH_2FCF_3 and the proper chemical name 1,1,1,2-tetrafluoroethane.

Freon 41	*Fluoromethane*	
R-11	CFC-11	Trichlorofluoromethane
R-12	CFC-12	Dichlorodifluoromethane
R-13	CFC-13	Chlorotrifluoromethane
R-22	HCFC-22	Chlorodifluoromethane
R-23	HFC-23	Trifluoromethane
R-113	CFC-113	Trichlorotrifluoroethane
R-114	CFC-114	1,2-dichloro-1,1,2,2-Tetrafluoroethane
R-115	CFC-115	1-chloro-1,1,2,2,2-Pentafluoroethane
R-116	CFC-116	Hexachloroethane
R-134a	HFC-134a	1,1,1,2-tetrafluoroethane
R-227ea	HFC-227ea	1,1,1,2,3,3,3-heptafluoropropane

Freon in copper tubing open to the environment can turn into phosgene gas after coming in contact with extreme heat, such as, while brazing or in a fire situation. Phosgene is a substance that was used as a chemical weapon in World War I. Low exposure can cause irritation, but high levels cause fluid to collect in the lungs, possibly resulting in death.

HALON

Halons are a group of compounds obtained by replacing the hydrogen atoms of a hydrocarbon with halogen atoms, such as bromine or fluorine. Halon 1211 is bromochlo-rodifluoromethane (CF_2BrCl) and Halon 1301 is bromotrifl-uoromethane (CF_3Br). Halons are very stable and are widely used in fire extinguishers where water and other alternatives would be ineffective and dangerous (*e.g.* when dealing with oil fires) or cause unacceptable collateral damage (*e.g.* with electronic equipment.)

Introduction

At high temperatures, halons decompose to release halogen atoms that combine readily with active hydrogen atoms, quenching the flame propagation reaction even when adequate fuel, oxygen and heat remains. The chemical reaction in a flame proceeds as a free radical chain reaction; by sequestering the radicals which propagate the reaction, halons are able to "poison" the fire at much lower concentrations than are required by fire suppressants using the more traditional methods of cooling, oxygen deprivation, or fuel dilution. For example, Halon 1301 flood systems are typically used at concentrations no higher than 6 per cent v/v in air, and can suppress many fires at 2.9 per cent v/v. By contrast, carbon dioxide fire suppression flood systems are operated from 34 per cent (surface-only combustion of liquid fuels) up to 75 per cent (dust traps). Carbon dioxide can cause severe distress at concentrations of 3 to 6 per cent, and has caused death by respiratory paralysis in a few minutes at 10 per cent concentration. Halon 1301 causes only slight giddiness at its effective concentration of 5 per cent, and even at 15 per cent persons remain conscious but impaired, and suffer no long term effects. (Experimental animals have also been exposed to 2 per cent concentrations of Halon 1301 for 30 hours per week for 4 months, with no discernible health effects at all.) Halon 1211 also has low toxicity, although it is more toxic than Halon 1301, and thus considered unsuitable for flooding systems.

Fig. Halon Canisters Used in a Fire-suppression System

However, halon fire suppression is not completely non-toxic; very high temperature flame, or contact with red-hot metal, can cause decomposition to toxic byproducts. The presence of such byproducts is readily detected because they include hydrobromic acid and hydrofluoric acid, which are intensely irritating. Halons are very effective on Class A (organic solids), B (flammable liquids and gases) and C (electrical) fires, but they are totally unsuitable for Class D (metal) fires, as they will not only produce toxic gas and fail to halt the fire, but in some cases pose a risk of explosion. Halons can be used on Class K (kitchen oils and greases) fires, but offer no advantages over specialised foams.

Nomenclature

Halons are referred to by a system of code numbers similar to (but simpler than) the system used for freons. The first digit specifies the number of carbon atoms in the molecule, the second is the number of fluorine atoms, the third is the chlorine atoms, and the fourth is the number of bromine atoms. If the number includes a fifth digit, the fifth number indicates the number of iodine atoms (though iodine in Halon is rare). (Note that the various atomic species are listed in order of increasing atomic number.) Any bonds not taken up by halogen atoms are then allocated to hydrogen atoms.

For example, consider Halon 1211: C F Cl Br 1 2 1 1

Halon 1211 has one carbon atom, two fluorine atoms, one chlorine atom and one bromine atom. A single carbon only has four bonds, all of which are taken by the halogen atoms, so there is no hydrogen. Thus its formula is CF_2BrCl, and its IUPAC name is therefore bromochlorodi fluoromethane.

Application

Halon 1211 is typically used in hand-held extinguishers, in which a stream of liquid halon is directed at a smaller fire by a user. The stream evaporates under reduced pressure, producing strong local cooling, as well as a high concentration of halon in the immediate vicinity of the fire. In this mode, extinguishment is achieved by cooling and oxygen deprivation at the core of the fire, as well as radical quenching over a larger area. After fire suppression, the halon moves away with the surrounding air, leaving no residue.

Halon 1301 is more usually employed in flooding systems. In these systems, banks of halon cylinders are kept pressurised to about 4 MPa with compressed nitrogen, and ducts lead to the protected enclosure. On triggering, the entire measured contents of one or more cylinders are dumped into the enclosure in a few seconds, through nozzles designed to ensure uniform mixing throughout the room. The quantity dumped is pre-calculated to achieve the desired concentration, typically 3-5 per cent v/v. This level is maintained for some time, typically twenty minutes, to ensure all items have cooled so reignition is unlikely to occur, then the air in the enclosure is purged. During this time the enclosure may be entered by persons wearing SCBA. (There exists a common myth that this is because halon is highly toxic; in fact it is because it can cause giddiness and mildly impaired perception, and also due to the risk of combustion byproducts.) Flooding systems may be manually operated or automatically triggered by a VESDA. In the latter case, a warning siren and strobe lamp will first be activated for a few seconds to warn personnel to evacuate the area. The rapid dumping of halon and consequent rapid cooling fills the air with fog, and is accompanied by a loud, disorienting noise.

Environmental Issues

There is concern that halons are being broken down in the atmosphere to bromine, which reacts with ozone, leading to depletion of the ozone layer (this is similar to the case of chlorofluorocarbons such as freon). These issues are complicated: the kinds of fires that require halon extinguishers to be put out will typically cause more damage to the ozone layer than the halon itself, not to mention human and property damage. However, fire extinguisher systems must be tested regularly, and these tests may lead to damage. As a result, some regulatory measures have been taken, and halons are being phased out in most of the world.

MOLECULE OF WATER

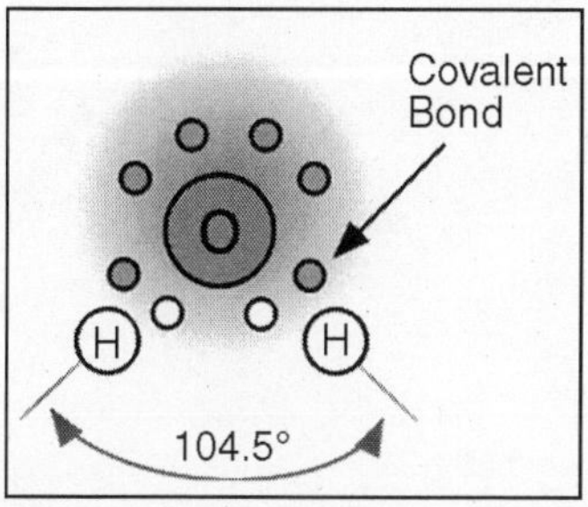

A molecule is an aggregation of atomic nuclei and electrons that is sufficiently stable to possess observable properties- and there are few molecules that are more stable and difficult to decompose than H_2O. In water, each hydrogen nucleus is bound to the central oxygen atom by a pair of electrons that are shared between them; chemists call this shared electron pair a covalent chemical bond. In H_2O, only two of the six outer-shell electrons of oxygen are used for this purpose, leaving four electrons which are organized into two non-bonding pairs. The four electron pairs surrounding the oxygen tend to arrange themselves as far from each other as possible in order to minimize repulsions between these clouds of negative charge.

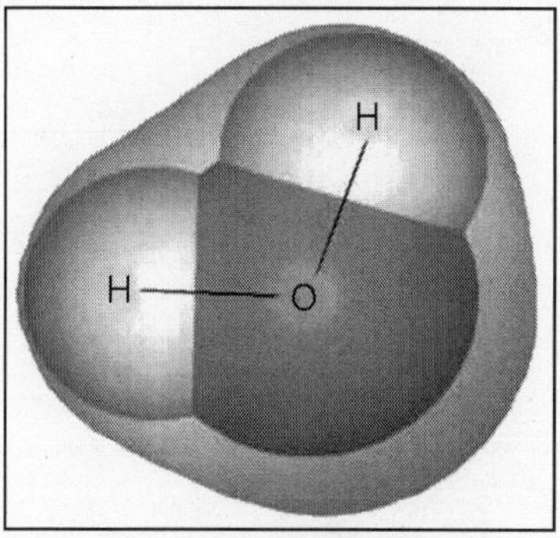

This would ordinarily result in a tetrahedral geometry in which the angle between electron pairs (and therefore the H-O-H bond angle) is 109.5°. However, because the two non-bonding pairs remain closer to the oxygen atom, these exert a stronger repulsion against the two covalent bonding pairs, effectively pushing the two hydrogen atoms closer together. The result is a distorted tetrahedral arrangement in which the H–O–H angle is 104.5°.

Because molecules are smaller than light waves, they cannot be observed directly, and must be "visualized"by alternative means. This computer-generated image comes from calculations that model the electron distribution in the H_2O molecule. The outer envelope shows the effective "surface"of the molecule as defined by the extent of the cloud of negative electric charge created by the ten electrons.

HYDROGEN BONDING

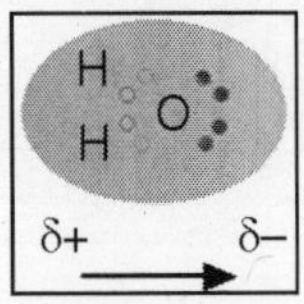

The H_2O molecule is electrically neutral, but the positive and negative charges are not distributed uniformly. This is illustrated by the gradation in Colour in the schematic diagram here. The electronic (negative) charge is concentrated at the oxygen end of the molecule, owing partly to the nonbonding electrons (solid blue circles), and to oxygen's high nuclear charge which exerts stronger attractions on the electrons. This charge displacement constitutes an *electric dipole,* represented by the arrow at the bottom; you can think of this dipole as the electrical "image" of a water molecule.

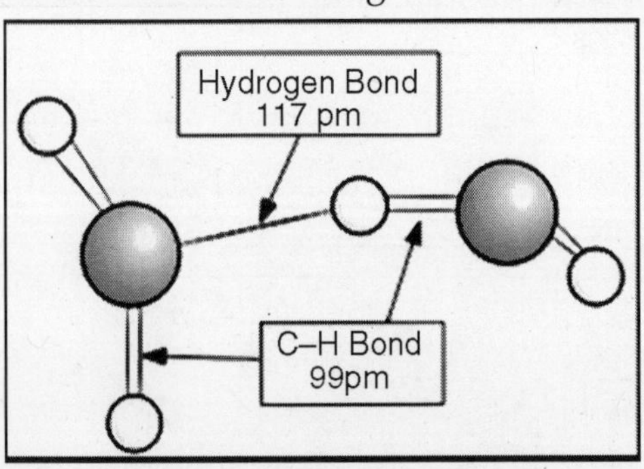

As we all learned in school, opposite charges attract, so the partially-positive hydrogen atom on one water molecule is electrostatically attracted to the partially-negative oxygen on a Neighbouring molecule. This process is called (somewhat misleadingly) *hydrogen bonding*.

Notice that the hydrogen bond (shown by the dashed green line) is somewhat longer than the covalent O—H bond. This means that it is considerably*weaker*; it is so weak, in fact,that a given hydrogen bond cannot survive for more than a tiny fraction of a second.

9

Environment and Agriculture Pollution Control System

WASTE MANAGEMENT AND POLLUTION CONTROL

Nowhere is the deterioration of the infrastructure for managing sewage and solid waste more evident than in Georgetown. Furthermore, run-off from agricultural activities discharges sediment and inputs, such as fertilisers and pesticides, into rivers and coastal waters, and is presumed to enter the defective water supply systems during flooding. Industrial activities in the coastal zone discharge untreated wastewater into water courses.

There is no monitoring of these impacts but the incidence of health problems suggests that improper management of wastes poses serious environmental and public health problems nationwide. Addressing these problems will be one of the main responsibilities of the new EPA.

LIQUID WASTE

Proper disposal of sewage is the primary liquid waste management issue. Georgetown is the only city served by a communal sewerage system. The main sewerage system, built between 1924 and 1929, services around 648 hectares in central Georgetown, containing approximately one-third of the city's population (about 80,000 people). The old gravity-piped sewerage system discharges to 24 pumping stations around the city. Power outages, aging pumps and solid waste dumping interfere with continuous flow through the system, which discharges an estimated 3.9 million gallons of untreated sewage per day through an outfall into the mouth of the Demerara River.

The University of Guyana and the Tucville area outside Georgetown also have sewer systems, but neither has an operating treatment plant at this time. The remaining population of greater Georgetown and the other major communities are served by septic tanks and pit latrines, which overflow from time to time to ditches and storm channels, causing foul odours, unsightly conditions and severe endemic mosquito problems. More serious is the potential for contaminated surface water from storm channels to enter drinking water mains at times of low or negative pressure and during flood conditions. In rural areas, septic tanks and pit latrines are commonly used for sewage disposal. Septic tanks employ a filter box for dispersing tank effluent into the soil to a depth that depends on the level of the water table. Pit latrines, if properly covered and maintained, can provide adequate sewage disposal for rural areas without a pipe-borne water supply.

Solid Waste

Solid waste disposal practices have not kept pace with the demands posed by increases in population and waste generation. Municipal solid waste management in Georgetown, more than elsewhere, suffers from years of under-funding. According to recent data, Georgetown generates approximately 50 tons of solid waste per day. Recently the city has had to use contractor vehicles for waste collection because of a lack of city garbage trucks. The contractor often interrupts the waste collection services because its fees are not paid for lack of funds.

A common practice throughout the city is the dumping of waste in nearby public areas and vacant lots rather than at official dump sites. The resulting waste pollution and open decomposition, and also drain clogging, is a common sight in all city districts. Georgetown has an old waste incinerator, used principally for disposal of hospital wastes, which is currently operating at only 10 per cent of capacity. Because of its location in the city and the costs of rehabilitation, it may be necessary to abandon the incinerator and simply use the site as a transfer station.

Otherwise, most of the city's solid waste is disposed in temporary, poorly designed landfills sited at various locations around the city. These makeshift urban landfills are shallow excavated trenches, backfilled with solid waste, and covered with soil by earthmoving equipment. Their current operation cause foul odours and problems of wind-blown waste. There are no controls on the waste that is dumped in the landfill and leachate from the landfill is believed to be entering the high water table underlying the city.

In the communities of Linden and New Amsterdam, municipal waste disposal is handled by open dumping and burning. No facility design or operating procedures are followed for these sites. Both air and ground-water pollution are cause for concern. However, no monitoring is currently done.

Agricultural Wastes

Animal wastes from pigs, cattle, and poultry are good for soil conditioning if properly managed. Nevertheless, they may cause pollution problems if run-off is allowed to enter streams and rivers, or to percolate into groundwater, as happens in the coastal zone. As livestock farms continue to develop, proper measures should be encouraged for the lagooning of animal waste, the use of solids to condition soils and of properly treated effluent to irrigate fields.

- *Industrial and Other Wastes*: Very little is known about industrial waste management. Currently no reliable data are available on the quantity, characteristics, and management practices for industrial liquid and solid wastes. Industrial solid wastes, which may include hazardous waste constituents, are generally trucked to the land disposal site in Georgetown. There is currently little monitoring or control over industrial waste disposal. GAHEF has recently begun to characterise industrial pollution on the assumption that disposal practices for industrial effluents pose environmental hazards and that proper management systems for these wastes must be introduced. GAHEF has recently assembled some information on industrial waste generated and the pollutants of potential concern.

Table. Sources of Industrial Waste in Guyana

Industry	Number	Potential Pollutants
Sawmills	66	BOD, dust
Food processing	47	BOD, phosphates, solids, dust, pathogens
Detergents/soaps	9	BOD, phosphates, caustics
Metalworking/foundry	8	Heavy metals, solids
Sugar refinery	7	BOD, solids, caustics, phosphates
Chemical/pharmaceutical	6	Acids, alkalies, phosphates, solids
Distilleries/breweries	5	BOD, phosphates, thermal
Plastics	4	CFCs, solids

Ship-generated wastes, which can be considerable in CARICOM islands with a strong tourism industry and regular visits from cruise lines, contribute minimally to waste in Guyana. The port in Georgetown handles mostly cargo boats and the solid waste is removed in bags and trucked to the Georgetown land disposal site. More of a concern is hospital waste (pathological, chemical, infectious, and pharmaceutical waste), which is currently bagged in plastic after separation and is collected and trucked to the Georgetown incinerator.

Environmental and Public Health Impacts

An assessment of the impacts of improper environmental and natural resource management practices reveals many areas that have potentially high negative impacts on human health and the environment. Current sewage disposal practices appear to cause faecal contamination of drinking water sources. Pollution of surface and ground water also has serious impacts on fisheries resources in coastal and marine waters, which then enters the food chain for the human population. In addition, the agricultural run-off which ultimately enters the coastal zone may contribute potentially significant pollutants in the form of increased biochemical oxygen demand (BOD) and nutrient enrichment. Such pollution may have serious impacts on aquatic and marine life and any contamination of drinking water from this run-off would impact human health. Untreated industrial effluents discharged into nearby canals and rivers will affect the quality of drinking water if not rapidly dissolved.

Lack of management of solid waste poses serious health threats. One primary concern is the increase in diseases carried by rats and vermin from solid waste dumps. Contamination of ground and surface water by leachate from solid waste dumps is likely. Where solid waste is burned as part of the disposal process, air pollution may pose a threat to the surrounding area.

Only limited water quality monitoring is done for drinking water sources (i.e., surface and ground waters) and no testing is done of rivers and coastal waters. Nevertheless, as noted, one significant indicator of the environmental health problem is the increase in the incidence of environmental diseases. Health data show that the population suffers from environment-related diseases, identified in Table, that are transmitted to humans from contaminated water, food, or soil.

Table. Environment-related Diseases

Transmission	Diseases
Water-borne	Cholera, dysentery, gastroenteritis, and typhoid
Food-borne	Dysentery, gastroenteritis, and infectious hepatitis
Soil-borne	Hookworm

Other diseases that afflict the population are transmitted by common vectors that are directly influenced by environmental conditions. These are shown in the following table.

Table. Vector Transmitted Diseases

Vector	Diseases
Mosquito	Dengue, filariasis, malaria, and yellow fever
Snail	Schistosomiasis

Recent health statistics show increases in the number of reported cases of malaria and a high mortality rate, especially in the interior. The increases in cases of gastroenteritis reflect poor conditions in water supply and sanitation, as does the outbreak of cholera along the border with Venezuela. In total, water-borne diseases are estimated to have risen more than fourfold over the last decade. In addition, the infant mortality rate is the highest in CARICOM.

PRIVATE SECTOR ENVIRONMENTAL IMPACTS

Because of both the increased public scrutiny of private sector developers and the environmental requirements of the international lending institutions, corporations that have obtained large forestry and mining concessions are increasingly required to introduce environmental protection standards in their operations. Currently, though, the capacity to monitor their compliance with these standards does not exist. Small scale private sector investors engage in gold mining but also in logging and sawmilling activities, as well as in agriculture and fishing, and there are few controls on their activities.

CULTURAL RESOURCES

The National Trust Act of 1972 legislates the preservation of historical and cultural areas. A National Trust Committee (NTC) has functioned sporadically since 1972. The NTC receives no funds from the Government and has no funds of its own. The NTC is assisted, though, by the Guyana Heritage Society (GHS) that has just completed a list of the material and immaterial heritage, some of which is in the putative national parks.

The unprioritised list comprises buildings in the city of Georgetown, items of archeological and zoological interest in the Rupununi, places of beauty, nesting places of birds, burial mounds, graves and domains of pioneers, churches and missions, and Amerindian burial urns. The fiscal crisis has prevented architectural conservation measures being enforced in Georgetown, which, if done, could enhance the city's value as a cultural asset and a tourist attraction.

M. PUBLIC AWARENESS

The level of public awareness about environmental issues has risen significantly in the wake of incidents such as the Omai cyanide spill. Sources for increasing public environmental awareness, i.e., public education and community outreach efforts by both Government and NGOs, have recently begun to be put in place.

The Environment Unit has a limited but active environmental education programme that functions under the supervision of a trained Environmental Officer, and it will be strengthened under the new Environmental Protection Agency. The programme has included a 20-minute television show, daily radio broadcasts on environmental issues, and two-day seminars on environmental awareness for primary and secondary school teachers. The Environment Unit has drawn up a five-year, multifaceted environmental education programme, including school curriculum changes, public outreach campaigns, etc., but it has not begun implementation for lack of financial resources. The Health Education Unit of the Georgetown City Council, helped by the Environment Unit, is also very active in this field, arranging lectures and competitions between schools, etc. The personnel working in this area is knowledgeable and doing the best it can with limited resources. The University of Guyana (UG) provides several environment-related courses within some of its programmes in the Physical Sciences. UG offers an Environmental Health Officers Diploma, a three-year programme that includes courses in environmental studies, epidemiology, public health, water, sewerage, and solid waste management. With support from the European Community, UG has recently established an Environmental Studies Unit, which offers a four-year undergraduate programme in environmental studies. Carried out jointly with the University of Utrecht in Holland and funded by the EEC, the new programme will establish physical facilities and provide a core of trained staff for UG to undertake teaching, environmental research, and consultancies.

Our environmental NGO movement is still in an embryonic stage and is only beginning to play a role in public education and advocacy on environmental issues. There are currently fewer than ten NGOs with an interest in environmental matters. Though limited in number, these NGOs have been involved mainly in the monitoring of the actions of companies in the areas of forestry and mining. The most common contribution of all the NGOs is in the area of public awareness and environmental education. Again, the Omai spill inadvertently gave these efforts a considerable boost.

POVERTY AND THE ENVIRONMENT

ENVIRONMENT AND LOW-INCOME GROUPS

An intensification of poverty in the whole society accompanied the protracted economic decline experienced in the 1970s and 1980s. Per capita income at US $686 in 1994 is one of the lowest in the Western hemisphere and had declined by almost one-third compared with 1980. Some studies on the extent of poverty put as much as 67 per cent to 75 per cent of the population below the poverty line in 1989. While there are issues concerning the methodology of these measurements, and the percentage in poverty has declined in recent years, poverty is indisputably a substantial problem.

A link between poverty and environment exists to some extent, despite the low population density. The most obvious examples are: the cutting of trees for charcoal and the reaping of mangroves for household use and cottage industry tanning; lack of sufficient sanitary facilities with the concomitant dangers for the purity of the aquifers; overfishing of inshore marine resources; and the environmental damage caused by the itinerant labour that characterises small-scale gold mining with its proliferation of sites. These activities have long-term destructive potential. Perhaps the saddest aspect of this linkage is that the environmental damage attributable to poverty in turn reduces further the level of well-being of the poor, by making their water less clean, their fuelwood more scarce, and their fish catches more sparse.

Environmental Consequences

The most visible poverty-environment link is to be found in the impact of the poverty of public sector institutions on environmental degradation. As is well known, a sharp decline in fiscal viability associated with the economic crisis led to a severe deterioration of the social and economic infrastructure, with grave consequences for environmental health, sustainable management of natural resources and protection of the environment.

The financial and manpower resource constraints at the Central, Regional and Local Government levels eroded, over time, the capacity to maintain and expand infrastructure. The evidence of collapse of the social and economic infrastructure has been everywhere — in the breaches in the sea defences and their consequences, the malfunctioning of the drainage and irrigation systems in coastal areas, the breakdown of the management of solid and liquid waste, the deterioration in water supply and electricity systems, the unavailability of affordable housing and consequent rise in squatting. In recent years the Government has started to come to effective grips with some of these problems, and this National Development Strategy is designed to maintain that forward momentum and broaden the economic and social advances.

Breakdown of the Social Infrastructure

The concentration of the population on the coastal strip below sea level means that most of the solid and liquid wastes of the country are generated in areas that are not well drained. Underfunding of the municipal authorities has resulted in a near total breakdown of solid waste management in Georgetown and in a lack of adequate maintenance of the drainage canals in the city (although the latter has begun to improve recently). The population's access to potable water has declined. Poor and crowded housing conditions characterise the more densely populated coastal belt. Unavailability of affordable housing for large numbers of the working population has led to squatter settlements along the embankment of established waterways, like the Lamaha canal, that convey water to the public water supply system.

One manifestation of the pervasive environmental problems is the extraordinary frequency of communicable diseases, with a high incidence of water and vector-borne diseases. Cholera, filaria, dengue, malaria, tuberculosis, gastroenteritis, typhoid, hepatitis and even cancer are largely environmental problems, some of them associated with pollution of the environment by chemical and bacteriological agents that is attendant on the nonexistence or the breakdown of potable water and waste disposal systems.

Poverty-related illnesses have been an increasing cause of death. The life expectancy declined from 70 years to 64.9 years during the period 1985-1992, which is about the same as that found in the late 1950s or early 1960s. Because of the deterioration of medical facilities and a serious shortage of staff, supplies and equipment, plus an increase in poverty levels over the past 28 years, levels of infant, child and maternal mortality rates are unacceptably high. However, it is important to note that life expectancy has ceased to decline and the national health services are reorganising themselves to better deal with some of these problems. What is now needed is a vigourous environmental policy to complement those efforts.

ENVIRONMENTAL FEATURES OF COASTAL ZONE

The Natural and Human Setting

The principal settled areas are uniquely dependent on sound environmental management. The coastal plain occupies approximately 7 per cent of the total area of the country and extends along the entire 430 km. of the Atlantic coast, varying in width from 26 to 77 km.

Five major wetland systems are distinguished in the coastal plain: the marine ecosystem of the sea coast, the estuarine ecosystems of tidal wetlands of the river mouths, and the riverain, palustrine and lacustrine ecosystems. The fertile plain consists of surface clays underlain by clays of the Demerara and Coropina formations. Extending for as much as five miles inland, much of the coastal plain lies between 0.5 metres and 1.0 metres below high tide levels. The shore zone consists of coastal works, mud banks, a mangrove belt and sand flats, all of which serve to protect the plain from flooding.

Ninety per cent of the estimated population of 751,000 resides in the coastal zone. Georgetown has a population of approximately 200,000 in its greater metropolitan area. Agriculture is still the major economic activity, accounting for 24 per cent of GDP and 35-40 per cent of employment. Except for forestry, all agricultural products (sugarcane, rice, other crops, livestock, and fishing) come from the coastal area. Most of the industry is found in Georgetown and along the remaining coast, with only a small concentration in Linden and north of Linden along the Demerara river. Other major economic activities and investments outside Georgetown and the coast consist mainly of large and small scale mining, mostly on major rivers. Groundwater provides 90 per cent of the potable water supply and is extracted mainly from the coastal artisan basin.

Complex interlinked erosion and flooding issues threaten the viability of the intensely concentrated economic base. Sea level rise and possible coastal subsidence add to the pressures on the sea defences and, inevitably, lead to greater coastal erosion and flooding. The shorezone, which serves as the natural line of defence against coastal erosion, is subject to erosion from floating mud shoals in the Atlantic that originate in the Amazon, exacerbated by destruction of mangroves and by removal of sand for construction on shore. The system of water conservancies and drainage and irrigation works in the coastal plain, which have also suffered from lack of maintenance, cause periodic fresh water flooding.

Coastal Erosion

Coastal erosion has been taking place at a relatively rapid rate. A comparison of the 1783 and 1970 coastlines around Liliendall shows a regression of almost 1 kilometre (erosion rates of -4 to -5 metres annually). Evidence of this retreat can be seen all along the coast — old sluice gates form isles far out to sea, and old shorelines and sand ridges run parallel to the present shoreline. Coastal erosion has many causes, some natural and some man-made, and distinguishing between the effects of each is difficult given the paucity of baseline data.

Mud shoals have had and continue to have an unquantifiable impact on coastal erosion. These shoals move along the Brazil, Suriname, Guyana coast from east to west in a series of waves and macro ripples at an approximate speed of 1.3 km per year. They are in constant motion and the refraction and concentration of wave energy between ripples greatly increases the capacity of the shoals to erode the coastal area. On the eastern side of the shoal, the coast accumulates the mud that is colonized by mangroves. However, as the mud shoal moves westward, the mangroves are unable to survive the wave attack and erosion takes place.

Mangroves are a first line of defence against wave action and storms on the coast and also provide a habitat for juvenile fish and shrimp. The construction of drainage systems for agricultural production projects in the coastal zone may have adversely impacted mangroves, as was evident once the MMA-ADA drainage and irrigation scheme was in operation. Mangroves are also plundered for fuelwood and for use in construction and tanning. No clear picture of the extent of the degradation of the mangroves is available, but prudence dictates that they should be protected from further destruction to help prevent further erosion of the coastline and also to preserve the breeding grounds for some marine species.

Sea level rise is another possible cause of coastal erosion. Tide gauge readings in Georgetown from 1960 to 1981 showed a rise of 9 cm. in the relative level of the sea. Engineers in the Hydraulics Division have observed water levels reaching higher up the sea walls. The observed rise in water levels, however, may be partly attributable to subsidence of the coastal zone, caused by the extraction of fresh water from the coastal aquifers. At present the data are inadequate to determine how much of the observed water level rise might be due to subsidence and how much is due to actual sea level rise.

Flooding and its Costs

The coastal plain is subject to flooding from both sea water intrusion and from fresh water overflows. Flooding destroys the value of the country's physical capital and its productivity, e.g., agricultural lands normally remain out of production for at least a year once flooded by saline water. Floods also have serious public health consequences in coastal areas. In Georgetown, sewage is discharged untreated through an outfall into the mouth of the Demerara river and during high tides and flooding this sewage is returned inland. Other areas use septic tanks and pit latrines, which may contribute to pollution of groundwater and, during flooding, surface waters.

For the past two decades, lack of maintenance, mainly because of budgetary constraints, has seriously weakened the sea defences. In 1991 more than 25 breaches of the sea wall flooded agricultural and residential areas. One breach at Cornelia on Leguan Island resulted in more than 1,000 acres of rice land being flooded. Since 1993 responsibility for maintenance of the sea defences was transferred from regional authorities to the Hydraulics Division, but the pervasive lack of funding and staff affects this Division too, and maintenance of these critical defence works is not yet at the desired standard. However, a new maintenance and rehabilitation programme is in place and so this concern should abate in coming years. The problem of contamination of groundwater, however, is still not being addressed systematically.

An intricate system of water conservancies and drainage and irrigation has been constructed to control flooding and overcome the seasonality of rainfall for coastal agriculture. However, for lack of maintenance, drainage ditches are often silted up or clogged with garbage, the mechanical pumps fail and sluice gates that control the flow of water often do not function properly, leading to frequent fresh water flooding. At high tides the flooding is exacerbated because gravity drainage seaward has to be blocked. The drainage and irrigation systems have profoundly altered the natural surface water regime, which is likely to have significant environmental repercussions. Nevertheless, no structured monitoring of such effects is conducted, except in the MMA-ADA Project.

Integrated Coastal Zone Management

Even though the coastal zone supports the majority of the population and the main economic activities, there is no formal coastal zone management plan or strategy for the multiple uses to which these resources may be put. Many agencies have shorezone management responsibilities, including the Environment Unit, the Hydraulics Division, the Hydro-Meteorology Department and Fisheries Department, and the Central Planning and Housing Department. This last agency is responsible for planning and zoning in urban and rural settlements, the implementation of standards and regulations and development control.

Government has recognised that improved coastal zone management depends on an integrated approach. In 1991 Government established a subcommittee on coastal zone management, formed by representatives from the major agencies dealing with the coastal zone. The next step should be more complete incorporation of coastal communities into the process of developing plans for the coastal zone and implementing them.

AREA OF ENVIRONMENTAL PROTECTION AGENCY

One of the first and most basic steps the Government will take in the area of environmental policy is the creation of an Environmental Protection Agency (EPA).

Its full functioning will entail the following steps:

- Create an Environmental Protection Agency (EPA). Through the mechanism of environmental coordination units, several governmental agencies will collabourate with the EPA in analysing issues, formulating solutions, implementing those solutions, and monitoring the results.
- Ensure that development of solutions for contamination problems will be preceded in each case by a detailed assessment of the problem and its origins, with quantification of the extent of the contamination to provide a baseline for measuring improvements.
- The EPA will be empowered to review plans and programmes and formulate recommendations for the Cabinet about the management of renewable natural resources with the objective of avoiding further resource degradation.

  Such plans and programmes include, but are not limited to, the following:
  - Forestry policies for sustainable harvesting of timber.
  - Fisheries management programmes.
  - Programmes for designating and managing protected areas and habitats.
  - Programmes and regulations governing the wildlife trade.
  - Policies and programmes for integrated coastal zone management.
- The EPA will also develop environmental education and awareness policies and programmes for the general public that focus on specific environmental issues and their solutions, in recognition of the fundamental fact that the public must be a partner in seeking and promoting effective solutions to environmental concerns.
- Environmental Impact Assessments (EIAs) will be required for any project or activity, in both the private and public sector, that significantly affects the environment.
- The EPA will maintain a staff of scientific experts who will carry out technical assessments of environmental problems and, when necessary, will testify on scientific issues before the courts. Legislation will be developed that recognises the role of scientific testimony in the court system; judges will apply specified criteria in the certification of experts who may make such testimony, whether they are from the EPA or elsewhere.
- Prior to the application of sanctions for industrial and agricultural pollution, remedial plans will be drawn up by the EPA and polluters

and the interested public will be given opportunities to file formal comments on them in draft. Once the plans are approved, they will have the force of regulations and sanctions will be applied for failure to comply with them.

By establishing the EPA and strengthening the environmental management capacity of selected sectoral agencies, ministries and local governments with significant environmental management responsibilities, it is expected that overall improvement in environmental management systems and capacity will be achieved.

This will only be possible by:

- Developing the human resource capacity to implement and monitor environmental management measures;
- Creating and strengthening the environmental management infrastructure;
- Streamlining and improving the legal and regulatory framework for environmental management; and
- Supporting national environmental education and public awareness programmes.

COASTAL ZONE MANAGEMENT

Successful identification, evaluation and mitigation and management of the causes and impacts of coastal zone degradation, depend on availability of baseline data, adequate monitoring and regulation by appropriate institutions designed for integrated, cross-sectoral management, complemented by appropriate and effective legislation, public awareness and participation by local communities in decision making and implementation.

Many of the issues relating to coastal erosion and land degradation in the coastal zone are complex, e.g., repair of the sea defences and of primary drainage and irrigation infrastructure, increasing the recharge of coastal aquifers, understanding the impacts and attempting to control the actions of the mud shoals, and monitoring sea level rise. Several of these and some less intractable issues are discussed below.

Institutional and Participatory Issues

The dispersion of responsibilities among the many institutions with responsibilities in the coastal zone makes integrated management difficult to implement, although such integration is essential for the effective management of such a complex area that is also so critical for economic development. Previous attempts to coordinate the activities of several agencies, through the Inter-Agency Committee on Environment and Development and the Advisory Environmental Council, were not successful. The Government is considering establishing a lead environmental agency with authority to coordinate agency activities and review private sector actions in the coastal zone. Involvement of coastal communities in developing and implementing solutions that will affect their well-being and that require their cooperation to be effective is of vital importance.

Infrastructure Maintenance

- *Sea Defences*: Sea defences are a fundamental element in the protection of the environment, protecting the coastal lands from flooding. The maintenance of sea defences remains a critical issue for the development of the agricultural sector, and indeed for the development of Guyana. Furthermore, the system of sea defence management should be able to react quickly to control breaches and floods in case of emergencies.
- *Primary D&I infrastructure*: The recently reconstituted national Drainage and Irrigation Board will take the responsibility for the maintenance of primary drainage and irrigation (D&I) works. As with sea defences, significant funds have been committed to the rehabilitation of the D&I system and maintenance of the system will be placed increasingly in the hands of users' associations.

Mangrove Management

Local control may be highly beneficial for mangrove management. Solutions for managing mangrove resources in a sustainable way will be developed and implemented with the participation of the local communities who exploit the mangroves for fuelwood. Alternative sources of firewood and a programme to replant mangroves need to be initiated in order to protect the mangroves, which are an important breeding ground for marine species that are caught by artisanal fishermen.

Other Priorities

Government established a prioritised list of actions, in the NEAP, for activities for which it requires donor assistance. *Excluding the activities discussed above, they are*:

- Initiating a study of the movement of the mud shoals;
- Initiating a study of the groundwater resources of coastal aquifers.

FORESTRY

To consolidate earlier initiatives, a Draft National Forestry Code of Practice has been formulated. This seeks to address many of the known issues and set out guidelines for sustainable and environmentally acceptable forestry practices.

To prevent loss of top soil and consequent fragmentation of forests, the specifications of individual FMPs, particularly as they relate to buffers, road plans, logging practices, etc., must be strictly adhered to in future.

Measures must be taken immediately to protect waterways and terrestrial areas from improper disposal of waste from logging camps and sawmills, by:

- Encouraging the use of sawmill waste for energy, by implementing NFAP project # 34 as a pilot project and sharing the results with the private sector;
- Designating approved sites for waste disposal at camp sites;
- prohibiting the dumping of camp waste in waterways.

Once the criteria and procedures for environmental impact assessments (EIAs) are in place, all forestry operations must be subjected to EIAs before and during the logging operation, under guidelines set and approved by the EPA. Currently, an area of 3.6 million hectares of forested land is accessible for exploitation and approximately 2.4 million hectares have been allocated for harvesting. The sector, however, is poised for an expansion of production. The potentially exploitable forests, for which access has to be constructed, extend over 10.4 million acres in the rain forest and in seasonal and in dry evergreen forests. Some 6.4 million acres, about 65 per cent of this total area, are currently leased for exploitation.

The Government is aware that the scale of the forestry exploitation that is currently envisaged will need to be carefully designed, regulated and monitored if Guyana is to avoid the plight of so many countries where the rainforest has been degraded and even destroyed for questionable short-term benefits. Furthermore, the emerging demand for "green seal" timber, i.e., wood and wood products that are certifiably produced in sustainable ways, is beginning to influence demand in international markets for forestry products. In Guyana, to date there has been very little capture of the economic benefits from non-timber products and services that traditionally include medicinal products, fibres and fruits and ecotourism. More recently, developed countries, through the GEF and the "Joint Implementation" provision of the Climate Convention, have made transfers to developing countries to conserve forest cover, particularly to maintain stored carbon in biomass to prevent it escaping into the atmosphere as carbon dioxide, one of the greenhouse gases.

Other possible funding sources for forest conservation are debt-for-development programmes, green venture capital, green investment funds and, perhaps most significantly, a special foundation that would mobilise donations from abroad to pay for non-timber concessions. Given these considerations, Government decided in May 1995 that no new logging concessions would be issued for a minimum of three years, or until the GFC is able to regulate forestry operations. Strengthening of the technical and monitoring capacity of the GFC is being carried out with assistance from ODA. Forestry regulations and the level of royalties are also being revised, in order to ensure that Guyana begins to be adequately compensated for the logging concessions it issues. On average, royalties and acreage fees have fallen by about 80 per cent in real terms since 1979 and require a five-fold increase to bring them back to those levels. Forest fees paid by loggers in Guyana are currently among the lowest internationally, even after allowance for the relatively lower quality and productivity in commercial timber species.

Conservation of the forest benefits the world at large, since, as stated earlier, Guyana is only one of thirteen countries with extensive rainforest left. Moreover, "the Guyana Shield", of which Guyana's forests are a part, is a unique and endangered region that stretches from Amapa in North-East Brazil, through French Guiana, Surinam, Guyana, Venezuela and Colombia until it reaches the Andes. The region is known to contain tremendous and largely endemic biological diversity. Although there is no comprehensive listing of medicinal plant species, Amerindian customs and practices indicate that as much as 10 per cent of the plant species may have medicinal properties. The forest also provides a habitat for wildlife and is of international significance for scientific and tropical forestry research. Furthermore, Guyana's forests make an essential contribution to the hydrological balance and climatic stability of the immediate region and the rest of the world.

Given the fiscal constraints the country faces in its quest for economic improvement, and the fact that conservation efforts bring benefits not only to Guyana but also to the world at large, mechanisms are required that will finance non-timber uses of the forests. Government will encourage efforts to establish a special foundation, which may be known tentatively as the Guyana Rainforest Foundation, that will mobilise funding from international NGOs, corporations, and bilateral governmental donors. This foundation would set up an endowment fund to receive the donations and then apply the earnings from the endowment to the payment of royalties for non-timber concessions, to the management of those concessions including the training and hiring of Amerindians and other local personnel as field staff and, if desired, to collabourate in the efforts to upgrade GFC's capacity to monitor timber concessions.

Royalties paid to the Government for non-timber concessions would be set at levels per acre that are comparable to timber royalties, thus compensating Guyana for the decision to forego the exploitation of some of its resources. The Government would expect this kind of initiative to be accompanied by increased lending from multilateral institutions, to generate additional income and employment in other sectors as offsets to the possible reductions of potential income and employment in forestry. The proposed Guyana Rainforest Foundation also would seek to promote ecotourism, medicinal uses of the forest, and other income-generating approaches. It would also promote international agreements on carbon offset (for industrial pollution in developed countries), as another source of compensation to Guyana for setting aside part of its natural resource base. There have been similar though not identical efforts of this nature in several countries in the

Hemisphere, of which the most successful example is Panama's Fundacion Natura, which has an endowment of US$25 million. If the international response is appropriate and Guyana can launch this initiative, it would become a model for many tropical countries. Government will do all in its power to make the foundation a success if a response is forthcoming from the international community.

BIODIVERSITY, NATIONAL PARKS AND PROTECTED AREAS

The World Wildlife Fund (WWF) developed a proposal in 1991 for expansion and management of the Kaieteur National Park. At its current size of 45 square miles, Kaieteur is much too small for the sustainable management of its biological diversity and high endemism. However, the greatest threat to Kaieteur is caused by miners who operate dredges in the Potaro river upstream from the falls and who live in a settlement above the falls. These mining operations and their settlement will have to be terminated and the recurrence of similar activities prevented. If they have a legal basis for their presence there, mechanisms will be found to compensate them for leaving.

The WWF proposed that Kaieteur National Park should include the entire upper watershed of the Potaro river with all its tributaries, and all forest lands bordering the savannahs to the southwest, the Kurungiku mountains to the southeast, Ebini mountain to the east, the Ayanganna mountain to the northwest and the upper watersheds of the Kuribrong and Amaila rivers. The inclusion of all the upper watersheds will ensure good water supply for Amerindian and other communities in the generally degraded savanna area to the west of the Ayanganna and Wokomung mountains.

The proposed area covers an estimated 400,000 to 450,000 hectares of some of Guyana's most diversified life zones, with one of the highest levels of endemic species found anywhere in South America. This area is relatively inaccessible because of the mountainous terrain. Two Amerindian villages fall within the proposed expanded area. The proposed boundaries would maintain the major ecosystem in the greatly enlarged area. The Government has decided to embark on the expansion of the Kaieteur National Park from its present size of 45 square miles to 222 square miles, which would incorporate the watershed leading into the falls. Government is currently reviewing the preliminary and coincident conclusions of studies recommending that the Kanuku mountains should be developed as a second National Park.

In 1994, the Government requested Global Environment Facility (GEF) funding for a project to develop a system of protected areas. According to the draft "Project Information Document" (PID) prepared by the Guyana Natural Resources Agency (GNRA) in August 1994, the principal objective of this project is to establish, according to internationally agreed criteria, a national system of protected areas that is fully representative of the country's varied and important biological diversity. This will entail undertaking thorough resource inventories and mapping to identify appropriate areas for designation; passing legislation to designate parks and reserves and to establish a Parks Management Agency, with authority for parks planning and management; beginning implementation of management plans in high-priority areas; and creating a park management training and environmental education programme. A project design workshop was held in November 1994, at which Government representatives, university staff, NGOs, and representatives of the GEF and other donors, agreed on the broad outline for the project.

Biodiversity is recognised to be an environmental asset and programmes will be designed for its protection and management, including but not limited to the above-mentioned protected areas schemes. The establishment of a system of protected areas is an essential step towards the sustainable management of Guyana's natural resources and the potential exploitation of its biodiversity resources. In addition, it an important step in attracting external funding for preservation of forested areas through the Joint Implementation provisions of the Climate Convention and directly from private sector energy corporations requiring carbon offsets and from non-profit organisations willing to invest in protection of bio-diversity. It would also enhance the possibilities of creating the proposed Rainforest Foundation. Effective management of these protected areas will require substantial initial technical assistance, funding for training and for the few capital investments required to establish the protected areas.

WILDLIFE

The Wildlife Services Unit, which regulates the wildlife trade, presently is poorly equipped to carry out this responsibility effectively. The monitoring and enforcement capabilities of the Wildlife Services Unit, as also the qualifications of its personnel, need to be strengthened. Its current mandate is only concerned with the harvest and export of wildlife and not with sustainable management of wildlife. To ensure that wildlife harvests do not exceed sustainable levels, closed seasons should be established for commercial species, with closed and open seasons for other species, together with appropriate export quotas which should be reviewed annually. Studies need to be undertaken to determine population levels and breeding habits and seasons and Guyana will need funds and technical assistance to undertake this work. To cater for the time frame within which accurate population and harvest rates can be determined, interim conservation harvests need to be established. The Government will initiate collabourative research arrangements and seek novel ways to attract funding for population studies. The updating and enactment of the Government's Conservation of Wildlife Bill, together with implementing regulations, are needed.

MINING

Special emphasis will be placed on working with GGMC and the private sector in developing appropriate technologies and incentives for solving problems of degradation of waterways and pollution of surface and groundwater sources stemming from mining activities. To this end Government will develop standards for levels of substances to be discharged and conditions for their discharge. There should also be a sampling and monitoring routine for potential contaminants. Any mining method or equipment which is damaging to the environment should be prohibited after an appropriate transition period.

To reduce the environmental health issues related to the use and discharge of potentially dangerous substances and to mining activities that create the conditions conducive to the spread of diseases and disease vectors, emission standards will be developed and implemented and the activities monitored. The costs of mitigatory measures should be borne by the polluter. Similarly, to reduce the levels of ambient dust and particulate matter released as a result of mineral mining, processing and transshipment, all mining operations with the potential to increase such air pollutant levels should be required to install equipment and implement measures to reduce such levels.

Activities of small scale miners need to be regulated and monitored. In particular, ways will be sought to have gold ore and dredgings from small-scale and artisanal miners processed in regional mills where mercury pollution can be controlled. The higher recovery rate in such mills should provide an economic margin which can be shared with the miners, as incentives for them to cooperate in the scheme. Simultaneously, the GGMC should explore the possibilities of obtaining technical assistance on best practices used by counterpart small scale miners in highly regulated countries, like Canada.

Every mining operation will be required to present an EIA, that will need to be critically reviewed before approval of the activity. The technical capacity of the GGMC will need to be strengthened for the review of EIAs. Local people from the communities where the mining activities occur should be trained to assist the GGMC with basic facets of the monitoring programme once a concession has been issued. The EPA will collabourate with the GGMC in monitoring progress on these important policy measures.

LAND AND AGRICULTURE

Land issues

As the Government reviews its economic policies in several areas, there is ample scope to apply cost recovery and "user pays" principles that will raise revenues and encourage conservation and environmental protection. Pricing policies for the use of government lands and for extraction of resources can be improved (on Agricultural Land Policy). Lease rates are being moved to market levels in stages and lease conditions are being modified to make them more suitable as collateral and more attractive to investors.

If the farmer has a clear title to the land being farmed, this by itself will create incentives for developing sustainable agricultural systems. A similar effect can be expected under the new class of leases.

Other measures to be taken in this area include:

- Giving the Lands and Surveys Commission autonomy, enabling it to have some degree of self-financing and allowing it to offer higher salaries, plus further strengthening, which will allow faster processing of lease applications.
- Relaxing controls on rental rates of private land, so that there will be less temptation to leave the land idle.
- Instituting a system of land taxes, for fiscal revenue purposes and also to discourage the practice of sometimes leaving land idle.
- Facilitating the granting of freehold titles in selected areas where there has been long-run beneficial occupation of leasehold land.
- Amending legislation with a focus on removing constraints to registering property in the Land Registry System.

Secondary D&I Works

So as to ensure the efficient and sustainable use of water resources and protect the coastlands from flooding, pilot programmes are being designed and implemented to transfer responsibility for secondary D&I management to the Water Users' Associations. The expectation is that by increasing the sense of ownership for resources, this will stimulate a greater responsibility in their use. Increased participation in D&I management is also possible by divesting responsibility to local governments and, ultimately, to the local communities themselves.

If local communities are to be empowered by involving them increasingly in service provision and resource allocation, the Government must seriously consider granting substantially greater autonomy to local authorities, primarily in the area of revenue generation and expenditure. Policies for this purpose are presented in, on the system of regional and local government.

Several quasi-governmental, NGOs and other organisations operate at the community level. At present a plethora of legislation covers cooperatives, friendly societies, producer associations, etc., which must be reviewed to remove constraints to their operations. Government should provide support facilities at the community level for such organisations, such as a Community Development Officer.

Research

Agricultural research in Guyana is primarily publicly funded and the public sector needs to take the initiative in promoting technologies that are environmentally sensitive to local conditions. Numerous regional (CATIE and IICA) and international agencies (the CGIAR centres, in particular) are designed specifically to provide such technical assistance. To access this assistance will require strengthening of the National Agricultural Research Institution through training, in particular. Some work has already been begun on reducing chemical use in agricultural production in Guyana through integrated pest management trials.

Pesticides

The Pesticides and Toxic Chemicals Control Bill, provides for regulation of the importation, sale and use of pesticides and toxic chemicals. Extension workers need to be trained in environmentally sustainable cropping activities for paddy production. The EPA will have the power to establish a regulatory regime and institutional framework for pollution control. A labouratory facility with chemical analysis facilities is an important investment priority if pesticide use is to be properly monitored.

FISHERIES

Concerns about the sustainability of yields from fisheries stocks underlie the policies presented in and the National Fisheries Management and Development Plan. Those policies and the ten-year plan, and the associated investment projects, together aim to bring sustainable development to the fisheries sector, including aquaculture. The fundamental orientation of the policies is to promote a gradual shift in emphasis in the sector, so that the areas with the greatest potential are encouraged to expand at the same time that safeguards are put in place to ensure that important species are not overexploited and exhausted, which could deprive fishermen of their source of livelihood. Increases in user charges (licences) will help to control excessive exploitation of the resources. Aquaculture will be encouraged given its economic potential, together with monitoring to ensure that its environmental impacts are limited.

Ecotourism

The objective for ecotourism development is to ensure that a high quality tourism product is made available in each specific area, without allowing tourism to have negative impacts on the natural environment or to disturb the plant and animal life.

This will entail:

- Identifying areas suitable for tourism development;
- Undertaking scientific research to determine the carrying capacity of sites selected for tourism activities;
- Implementing monitoring measures to ensure that the limits of carrying capacities are being observed;

- In the absence of such limits, best-scenario decisions on arbitrary limits should be determined for each site and strictly enforced.

To ensure that accommodations for tourists interested in both nature and urban sites are in keeping with the environment, guidelines for the design and construction of tourist accommodation should be developed. Construction in the interior should be done to the extent possible with natural and indigenous materials. Zoning laws and their regulation should be developed and enforced.

While tourism should fully involve indigenous populations as participants and beneficiaries, their cultural identify must be preserved and protected and they should participate in developing policies to achieve these ends. Guyana's cultural heritage is regarded as part of the environment, and the EPA will work with the relevant bodies in the public and private sectors to promote its proper conservation and stewardship.

SOURCES OF AGRICULTURE POLLUTION

The primary agricultural nonpoint source pollutants are nutrients (particularly nitrogen and phosphorus), sediment, animal wastes, pesticides, and salts. Agricultural nonpoint sources enter srface water through direct surface run-off or through seepage to ground water that discharges to a surface water outlet. Various farming activities result in the erosion of soil particles. The sediment produced by erosion can damage fish habitat and wetlands and, in addition, often transports excess agricultural chemicals resulting in contaminated run-off.

This runnoff in turn affects changes to aquatic habitat such as temperature increases and decreased oxygen. The most common sources of excess nutrients in surface water from nonpoint sources are chemical fertilizers and manure from animal facilities. Such nutrients cause eutrophication in surface water. Pesticides used for pest control in agricultural operations can also contaminate surface as well as ground-water resources. Return flows, run-off, and leachate from irrigated lands may transport sediment, nutrients, salts, and other materials. Finally, improper grazing practices in riparian, as well as upland areas, can also cause water quality degradation.

MANAGEMENT MEASURES

Sediment/Erosion Control

Soil erosion is one of the leading causes of water pollution in the United States. The goal of this measure is to minimize the delivery of sediment from agricultural lands to receiving waters. Land owners have a choice of one of two approaches: (1) apply the erosion component of the U.S. Department of Agricultures Conservation Management System through such practices as conservation tillage, strip cropping, contour farming, and terracing OR (2) design and install a combination of practices to remove settleable solids and associated pollutants in run-off for all but the larger storms.

CONFINED ANIMAL FACILITY

Animal waste contaminates many of our waters with pathogens and nutrients. The management measure for ALL new facilities and existing facilities over a certain size is to limit discharges from confined animal facilities to waters of the United States by storing wastewater and run-off caused by all storms up to and including the 25-year, 24-hour frequency storm. For smaller existing facilities, the management measure is to design and implement systems that collect solids, reduce contaminant concentrations, and reduce run-off to minimize the discharge of contaminants in both facility wastewater and run-off caused by all storms up to and including 25-year, 24-hour frequency storms. This measure also specifies management of stored run-off and solids through proper waste utilization and use of disposal methods which minimize impacts to surface/ground water. Confined animal facilities required to obtain a discharge permit under the NPDES permit programme are not subject to these management measures.

NUTRIENT MANAGEMENT

This measure calls for development and implementation of comprehensive nutrient management plans. The fundamentals of a comprehensive nutrient management plan include a nutrient budget for the crop, identification of the types and amounts of nutrients necessary to produce a crop based on realistic crop yield expectations, and an identification of the environmental hazards of the site. Other items called for in the measure include soil tests and other tests to determine crop nutrient needs and proper calibration of nutrient equipment.

PESTICIDE MANAGEMENT

This measure is designed to minimize water quality problems by reducing pesticide use,improving the timing and efficiency of application, preventing backflow of pesticides into water supplies, and improving calibration of pesticide spray equipment. A key component of this measure is use of integrated pest management (IPM) strategies. IPM strategies include evaluating current pest problems in relation to the cropping history, previous pest control measures, and applying pesticides only when an economic benefit to the producer will be achieved, i.e., application based on economic thresholds. If pesticide applications are necessary, pesticides should be selected based on consideration of their environmental impacts such as persistence, toxicity, and leaching potential.

LIVESTOCK GRAZING

The goal of this measure is to protect sensitive areas. Sensitive areas include streambanks, wetlands, estuaries, ponds, lake shores, and riparian zones. Protection is to be achieved with improved grazing management that reduces the physical distance and direct loading of animal waste and sediment caused by livestock by restricting livestock access to sensitive areas through a range of options.

In addition, upland erosion is to be reduced by either:

- Applying the range and pasture components of a Conservation Management System or
- Maintaining the land in accordance with the activity plans established by either the Bureau of Land Management or the Forest Service. Such techniques include the restriction of livestock from sensitive areas through locating salt, shade, and alternative drinking sources away from sensitive areas, and providing livestock stream crossings.

IRRIGATION

This measure promotes an effective irrigation system that delivers necessary quantities of water yet reduces nonpoint pollution to surface waters and groundwater. To achieve this, the measure calls for uniform application of water based upon an accurate measurement of cropwater needs and the volume of irrigation water applied. When applying chemicals through irrigation (a process known as chemigation), special additional precautions apply. The measure also recognizes that states water laws that conflict with the measure will take precedence over the measure.

COASTAL NONPOINT POLLUTION PROGRAMME

Section 6217 of the Coastal Zone Act Reauthorization Amendments of 1990 (CZARA) requires coastal states with approved coastal zone management programmes to address nonpoint pollution impacting or threatening coastal waters. States must submit Coastal Nonpoint Pollution Control Programmes for approval to both the U.S. Environmental Protection Agency (EPA) and the National Oceanic and Atmospheric Administration (NOAA). Requirements for state programmes are described in a document entitled "Coastal Nonpoint Pollution Control Programme: Programme Development and Approval Guidance" and are summarized in a separate fact sheet.

MANAGEMENT MEASURES

CZARA requires EPA, in consultation with NOAA and other federal agencies, to publish guidance specifying management measures to restore and protect coastal waters from specific categories of nonpoint source pollution. EPA has done so in a document entitled "Guidance Specifying Management Measures for Sources of Nonpoint Pollution in Coastal Waters". State Coastal Nonpoint Programmes must provide for implementation of these measures or alternative management measures in conformity with these measures in the coastal management area generally. "Management measures" are defined by law to be economically achievable measures that reflect the best available technology for reducing pollutants.

States may select from a wide range of practices or combinations of practices that will achieve the level of control specified in the management measure. This fact sheet summarizes the management measures applicable to agricultural sources. Other fact sheets summarize the measures for forestry, urban areas, marinas and recreational boating, hydro- modification, and wetlands/riparian areas.

NUTRIENT CONTROL

The control of nutrients is an important issue throughout much of the world, both from a public health perspective and to keep natural waters free from eutrophication. The most widely used water quality standard for nitrate (NO3 -) is the 50 mg l-1 limit adopted by WHO as a precautionary level to safeguard babies from the risks of contracting methaemo-globinaemia. Most national authorities regard the 50 mg l-1 concentration as a realistic target in relation to eutrophication and, therefore, programmes aimed at controlling eutrophication often use this value as an EQS. Whereas there are process techniques available to remove nitrate from drinking water after it has been abstracted (allowing higher levels to be tolerated in raw water used for potable supply) the eutrophication problem is universally dependent on the control of nitrate sources. There is an added and unexpected health implication related to eutrophication, particularly in lakes. In a number of countries in the world, the phenomenon of excessive growth of blue-green algae (notably *Microcystis* spp.) has caused concern where recreational pursuits take place on the lakes and also where the water is withdrawn for public supply. This is due to the recognition that such algae produce a number of toxins which, if ingested, can cause liver damage.

There are a number of well documented cases where animals are known to have died as the result of drinking water heavily laden with these algae (e.g. Australia, South Africa, UK) and, for this reason alone, nutrient control is justified. The most common source of the nutrients nitrogen and phosphorus is agriculture, and this is closely followed in the industrialised world by sewage effluents. The reduction of nitrogen and phosphorus from agriculture relies upon changes to farming practices because they give rise to diffuse sources.

Nitrogen

Ploughing of grassland and other crops, particularly during autumn, leads to the release of large quantities of soil nitrogen and, therefore, a general move towards permanent pasture regimes assists in lowering nitrate leaching. When this is not possible, the use of short-term rotational crops to take up nitrogen, followed by their harvesting and subsequent removal from the catchment, is helpful. Animal wastes should be used carefully, avoiding over-use and direct run-off into water courses; but wherever possible they should be used in place of synthetic fertilisers. Use of all types of fertilisers should be carefully controlled and matched to crop requirements. In Europe, legal control exists through the Directive Concerning the Protection of Waters against Pollution caused by Nitrates from Agricultural Sources (91/676/EEC). The purpose of this Directive is to reduce and to prevent pollution of fresh surface water, estuarine and coastal waters which arises from diffuse sources of nitrates. Within two years, member states have to identify and designate vulnerable zones, i.e. all areas of land draining into waters affected by pollution and that contribute to the pollution.

Action plans must be presented to improve the situation in these zones by specifying periods when land application of fertilisers is prohibited, by quantifying criteria for land application rates and by limiting use according to codes of good agricultural practice. Under the Water Resources Act, the UK government has introduced a further measure to reduce nitrate pollution through the identification of nitrate-sensitive areas. These reflect a specific clause in the legislation, Section 92, which permits the identification of such areas, and allows compensation payments to farmers in exchange for a reduction in the amount of nitrogenous fertiliser used. Such areas have been associated mainly with individual groundwater zones where nitrate concentrations have exceeded the standard of 50 mg l-1. Farmers enter into such agreements on a voluntary basis, for a five year period, and are set limits on the amount and timing of fertiliser application. In addition, green crops have to be planted in winter to minimise nitrate leaching. The Agri-Environment Initiative of the EU is also a voluntary scheme introduced in 1994 to encourage farmers to reduce the use of nitrates. Those doing so receive compensatory payment.

Phosphorus

A key issue controlling phosphorus input from agriculture is the need to prevent erosion from field surfaces. Phosphate tends to bind to soil particles which, when washed from fields into watercourses, become a source of phosphate in suspended form and in deposited sediments. Sediments act as a long-term source of phosphate by releasing it (i.e. by redissolution) under certain environmental conditions. Physical removal of the sediment layer, in order to remove the bound phosphate from the catchment, has been tried in a number of locations around the world. Some success has been achieved in lowering phosphate levels in the Norfolk Broads in England by a combination of the diversion of effluents containing phosphorus out of the area, phosphorus stripping at sewage treatment works, and by the dredging of 1 m of sediment. Concentrations below the target of 100 ìg l-1 of phosphorus were reached.

Pesticides

Pesticides represent a particularly difficult area of pollution control activity, not only because the environmental effects in relation to aquatic flora and fauna are important but because human health issues have a very important bearing upon the nature of the controls applied. There are several thousands of formulations of insecticides, herbicides and fungicides in common use and, therefore, the potential for water pollution is very high. There are also very stringent limits for water used for public supply and, consequently, the control of pollution by pesticides is crucial in water supply catchments. Most pesticides in waters are derived from agricultural use and it is therefore difficult to regulate their input to water bodies. Regulation can only take place by prevention, i.e. by indirect controls on their manufacture, storage and use. Approval for the use of pesticides is granted in the UK by government, following expert assessment of safety and the environmental risks. Authorisation is harmonised throughout Europe by the Directive concerning the Placing of Plant Protection Products on the Market (91/414/EEC) in which uniform principles for the authorisation process are adopted by member states. The active ingredients in pesticides are approved by the European Union and placed on an approved list. This Directive allows authorisation provided the pesticide is not expected to occur in groundwater at concentrations above 0.1 ìg l-1. Further controls are also placed on products by Directives such as that on Classification, Packaging and Labelling of Dangerous Substances. Once a pesticide is in use, it is controlled by safety legislation, such as the UK Food and Environmental Protection Act of 1985. A large number of guidance manuals have been issued on the safety precautions to be taken. These manuals usually indicate pollution prevention precautions and include advice on storage, on the disposal of unused material and application. In addition to specific legislation, a number of pesticides appear in other EU directives and in reduction programmes agreed in international protocols such as the Declaration on the North Sea. Several have also been totally banned because of their environmental impacts.

URBAN SOURCES

The major sources of urban pollution are urban stormwater run-off discharged through road drains or combined sewer outfalls, industrial area drainage discharged through surface water drainage systems (including spills of chemicals and oil) and refuse or solid waste drainage.

Run-off from Roads

Urban pollution occurs largely as a result of run-off from roads. Road surfaces are generally impermeable and thus any polluting material falling on them is, eventually, washed into a receiving watercourse or finds its way into groundwater. Such pollutants arise from many sources, the most important of which are traffic and maintenance operations. Traffic generates pollutants from vehicle emissions, including volatile solids, polynuclear aromatic hydrocarbons derived from unburned fuel, lead compounds and hydrocarbons. On main roads, leaks from lubrication systems provide a continuous source of fluid hydrocarbons. Abrasion of tyres during normal wear releases zinc, lead and hydrocarbons. Research work in Germany has indicated that tyre abrasion on motorways can release typically 572 g ha-1 a-1 lead, 120 g ha-1 a-1 chromium and 115 g ha-1 a-1 nickel (Muschack, 1990). Corrosion of vehicles also contributes quantities of metals, including chromium and lead. Road maintenance, particularly de-icing, is an important source of pollution, e.g. salt and urea. The impurities in road-grade salt can contribute to water quality deterioration. Roadside weed control also leads to diffuse sources of pesticide pollution. In addition to the pollution arising directly from road use and maintenance, road drains accept pollutants from atmospheric deposition, agricultural activities (after heavy rainfall) and general littering. Animal wastes, rich in bacteria, can also accumulate and contribute to the high levels of micro-organisms found in some waters. Oil pollution associated with vehicle maintenance is a specific problem in many areas. It is impractical and virtually impossible to control the quantity or quality of road run-off by normal regulatory means. As a result it is necessary to rely on good design of drainage systems with adequate built-in protection and on maintenance procedures which minimise the risks of pollution.

Various studies have been carried out to determine the most appropriate measures, and these have resulted in the following guidance:

- Gully pots, filter drains and soak-aways, which are all commonly used, can assist the removal of sediment but, unless maintained properly, can also pose a threat to surface and groundwaters. Infiltration basins and trenches can remove suspended material and possibly some dissolved pollutants, but can also be a threat to water quality. Detention tanks, storage ponds and sedimentation tanks operated at the end of the drainage system are successful in removing sediments. Lagoons and purpose built wetlands are capable of treating many potential pollutants, largely through the action of the associated vegetation.
- Liaison between the regulator and the highway authority is essential to ensure that suitable systems are installed when road schemes and urban development is planned. In the UK, the Department of Transport has issued design guidance and codes of good practice for routine and winter maintenance which include information for pollution prevention.

Urban Pollution from Separate Drainage Systems

In towns and cities, drainage systems can be of two types, combined sewers or separate sewers. In the first case drainage from roads, rooftops and similar impermeable surfaces is accepted into the foul sewerage network for treatment at sewage works along with domestic sewage and industrial wastes. In the second case, domestic sewage and industrial waste is separated for treatment, and the wastewater arising from rainfall run-off is discharged directly to watercourses without treatment. Drains in heavily urbanised areas may accept a variety of pollutants from rooftops, lorry loading bays, industrial sites and even from illegal connections to the surface water system. Publicity and inspections are needed to ensure that unexpected pollution does not arise from these sources. Pollutant loads discharged from urban drainage systems vary depending on local rainfall patterns, the variety of materials entering the sewer network, and the processes of mixing and degradation that occur in the sewers. The impact of the pollutant load is also complicated by variations in flow and quality in the receiving watercourse. Careful planning of the sewerage network is required to address these problems. Various modelling tools, such as rainfall modelling, sewer and sewage treatment modelling and river quality modelling, are now available to assist in this planning process. A simplified model SIMPOL (Spreadsheet Simplified Urban Pollution Model) combining many of the key processes has also been developed to test rapidly the performance of potential solutions and to identify rainfall events which could lead to important impacts on river water quality.

Nutrient Control

A number of specific measures have been adopted to deal with nutrient problems from sewage effluents. Legal instruments relate primarily to the control of point sources of nitrate and phosphate from sewage works - the most recent international measure being the adoption of the EC Urban Waste-water Treatment Directive in 1991 (91/27 I/EEC). The nutrient control measures therein comprise a requirement to add tertiary treatment (nitrate and phosphate removal) to plants which discharge into "sensitive areas". These are defined by reference to the eutrophic state of the water receiving the effluent.

In the UK, it has been decided to concentrate on phosphate removal for treatment plants discharging to inland waters and nitrate removal for treatment plants where the effluents enter marine waters (representing the critical elements for eutrophication in the respective situations). Other countries, such as Germany, intend to add nutrient removal to all of its larger sewage works. In addition to agricultural sources, phosphate occurs in sewage effluents as the result of its use as a detergent "builder" (creating optimum conditions in the wash water for the surfactants to operate). Sodium tripolyphosphate is commonly used; this compound breaks down to orthophosphate which can be used a nutrient source by aquatic plants. Control of the phosphate at source is not yet practised and removal relies on "stripping" of the orthophosphate from the sewage works effluent by chemical treatment. Sewage works can remove about 40 per cent of incoming phosphate, but removal of at least 90 per cent is often required to bring about a change in the trophic state of a receiving water body. Reductions in the polyphosphate content of some detergents is also assisting in this process. The importance of phosphates is recognised by the EC Urban Wastewater Treatment Directive by the inclusion of phosphate limits for discharges to eutrophic waters (2 mg l^{-1} for populations between 10,000 and 100,000 and 1 mg l^{-1} for populations greater than 100,000). The criteria for recognising eutrophication have been defined in a publicly available document which includes proposed EQSs for phosphate in different waters. This guidance has been used to identify eutrophic waters throughout the UK and enables the regulators to indicate where phosphate stripping from sewage works effluent is justified.

CATCHMENT MANAGEMENT PLANNING

In several countries, including the UK, Canada and the USA, the process of catchment or watershed planning has been introduced. This has resulted in a recognition of the importance of non-point sources of pollution and of the need to manage it as carefully as point sources of pollution. Catchment management plans are designed to cover a catchment of a river and its tributaries and any associated groundwater flows.

A catchment is a discrete area of land which has a common drainage system. The surface water catchment is defined by the topography of the land although this may not coincide with any associated groundwater catchment (which is influenced by the underlying strata). The principles of catchment management planning, however, apply to both. In terms of the impact of activities within them, catchments are largely self-contained, manageable units, although such activities can affect downstream areas. Catchment management plans are designed to be strategic in approach and to take into account regional and national policies and all activities likely to have an impact on the watercourse. The key attributes of a properly prepared plan are set out in a consultation report for all interested parties (local industries, public bodies and the local population). Following the consultation stage a final plan is prepared for adoption.

The consultation report should contain:

- Records of physical attributes and catchment uses.
- Proposed environmental targets.
- Comparisons of targets with current status of the water environment.
- Identification of issues and options for addressing them.

Its preparation will involve undertaking consultations on the uses, targets, issues and options, the preparation of action plans to address the issues, implementing the actions, and monitoring and reviewing the plan. The final plan should contain a future vision for the catchment (having taken note of the results of the consultation exercises), an overview of the catchment and action plans. As a result of the plan the main polluting sources should be identified and solutions agreed which can be achieved with available resources and to a timescale agreed by all concerned.

AGRICULTURAL POLLUTION AFFECTING RIVER ON A LOCAL SCALE

There are high concentrations of nitrates in the local groundwater, especially to the north-east of Canterbury (in the Thanet area). This is a Nitrate Vulnerable Zone (NVZ), where pollution levels exceed 50 mg/litre. The EU Drinking Water Directive set this limit for safe nitrate levels in 1980. In 1991, the EU issued another Nitrate Directive which required countries to identify NVZs where excessive nitrate pollution came from agricultural sources. Thanet is one of 68 NVZs in the UK. Locally, farming intensity is not as high as in Thanet, but eutrophication is bad in some of the marsh dykes (Stour Marshes).

Excessive demand for water (farmers are licensed by the Environment Agency to abstract water directly from the river for spray irrigation) can result in low flows, reducing the dilution capacity of the river; this is worse in summer, and particularly in drought years. This is known as a dilution effect, i.e. concentration increases as flow decreases. In wet winters however, prolonged and heavy rainfall may result in organic pollution incidents, with increased run-off from fertilized fields and accidental overflow from slurry pits. This is known as a pumping effect, i.e. concentration increases as flow increases, flushing chemicals off the fields and farmyards. These two responses seem to work in opposite directions, making the evaluation of diffuse pollution quite difficult.

PREVENTING AGRICULTURAL POLLUTION

The Ministry of Agriculture, Food and Fisheries - MAFF produced in 1991 a "Code of Good Agricultural Practice" for farmers giving detailed advice on the treating, storing and applying of animal livestock waste, the disposal of dirty water, fertilizers, fuel oil, sheep dip, pesticides, nitrates, disposal of animal carcasses, etc. With EA advice also, local farmers all have to produce a Farm Waste Management Plan to comply with pollution control regulations.

Some examples of good practice are:

- Reducing ploughing in the autumn
- Delaying the ploughing in of crop residues
- Reducing the amount of fertilizers, manure and Sewage sludge applied
- Sowing autumn crops early
- Sowing cover crops in winter to avoid bare ground
- Careful management of disposal of farm waste

- Set-aside regulations will also have the beneficial effect of reducing intensity of farming in the local area, and thus lowering nitrate levels.

ORCHARDS

Fruit farming locally forms part of the North Kent Fruit Belt. Orchards, small fruits and hops are all grown in this area, taking advantage of the lighter sandy soils of the Tertiary rocks, sandwiched between the heavier clay soils to the north and the Chalk loams to the south,

POLLUTION THREATS FROM ORCHARDS

The potential threat to water pollution comes from the intensive use of pesticides. 'Pesticide' is a general term which includes herbicides, fungicides and insecticides. They are widely used for weed control in agriculture, but also on roadsides and railway embankments. Spray drift from pesticide application can enter water courses if orchards are located too close to the river. The main threat however comes, not from spraying, but from poor storage and accidental spillage, as a result of which pesticides may get into, and contaminate, the groundwater.

Pesticides are persistent in the food chain, since the chemicals involved are non-degradable; they are said to 'bioaccumulation' in the food chain. Organochlorides are found to cause changes in the sexual and reproductive characteristics of wildlife. Top carnivores (in the river, fish) are especially affected. Since pesticides are found in very low concentrations in water, their detection and measurement is complex and expensive. The maximum admissible concentration (MAC) is extremely small - 100 ng/litre or 1 part in 10,000,000 for any one individual substance, but 500 ng/litre for total pesticide residues.

ORCHARD POLLUTION AFFECT

Spray application has never been a problem locally.. Triazines and Drins are representative groups of pesticide compounds, minute traces of which have been detected in the Great Stour.

PREVENTING ORCHARD POLLUTION

Local farmers follow the MAFF "Code of Good Agricultural Practice" and are required to produce a "Farm Waste Management Plan" for the Environment Agency. Correct spraying techniques have to be observed, under safe weather conditions.

For more than a century air pollution has affected agriculture. Burning coal and petroleum produces poisonous sulfur oxides. Fluorides result from mining, smelting and glass and ceramic manufacture. Rising levels of ammonia, chlorine, ethylene, mercaptans, carbon monoxide, and nitrogen oxides are poisoning the air we breathe and systemically contaminate the global soil and water.

Motor vehicles and growing population increasing at more than 80 million every year, after deduction of all deaths, produce photochemical air pollution affecting not only the urban city concentrations but also the contiguous rural areas.

The mixture of pollutants from all sources, including agriculture, has released a host of contaminants into the air, such as aldehydes, hydrocarbons, organic acids, ozone at ground level while simultaneously depleting the protective pencil thin ozone layer in the stratosphere, peroxyacetal nitrates, pesticides, and radionuclides. The damaging effect of these pollutants on food, fibre, forage, and forest crops is reflected in the global increase in cancer, asthmatic, bronchial and respiratory diseases which affect ever more humans.

The effects of air pollution on plants and animals and consequentially humans may be measured by the following factors:

- Interference with enzyme systems
- Change in cellular chemical constituents and physical structure
- Retardation of growth and reduced production because of metabolic changes
- Acute, immediate tissue degeneration.

These effects are counteracted and thereby aggravated by ever larger inputs of pesticides, insecticides, fungicides, herbicides and chemical fertilisers that are now producing negative returns while increasing systemic pollution of the foodchain. The increasing demand for food due to the increasing global populations of humans and feedlot animals for meat has led to genetic engineering that spread harmful systemic pollution worldwide simultaneously causing extinction of valuable indigenous crops.

Harmful pollutants that enter the air from sources other than agriculture and that damage plants and crops are classified as:

- Acid gases
- Products of combustion
- Products of reactions in the air
- Miscellaneous effluents

ACID GASES

Acid gases include fluorides, sulfur dioxide, and chlorine. Hydrogen fluoride is extremely toxic to plants; some plants are injured by contact with concentrations of less than one part per billion. Sulfur dioxide given off in combustion of oil and coal commonly causes necrosis (cell death).

PRODUCTS OF COMBUSTION

The primary products of combustion are ethylene, acetylene, propylene and carbon monoxide. Ethylene which is the most damaging to plants is derived mostly from motor vehicle exhausts and chemical industries such as polyethylene factories. Ethylene, ozone and peroxyacetyl nitrate are produced as reaction products in the air and are clearly implicated in plant injury. Ozone is a major toxic air pollutant affecting agriculture. This is not to be confused with the pencil thin ozone layer in the stratosphere that protects all life from the harmful effects of solar radiation.

POLLUTION BY AGRICULTURE

Agriculture pollutes the soil, air and water through pesticides, odours, methane, smoke, dust, allergenic pollens, and waste. The systemic damage to the global air, water and soil through genetically engineered crops cannot as yet be quantified. It could lead to globally changed conditions for all life - the consequences of which could be irreversible and fatal.

Eutrophication

Reduced levels of dissolved oxygen in water due to increased mineral and organic nutrient deposits produces algae and other water plants that choke other forms of life in the oxygen competition. Soil erosion, phosphorous and direct run-off from feedlot operations and intensive agriculture are the main cause.

Salinity

Through evapotranspiration, salts in irrigation water become more concentrated in the drainage affluent; the soil accumulates the residual salt which is added to what the arid soils already have in excess. It is further aggravated by the attempted removal of excessive salts through leaching causing further soil erosion.

Agricultural Processing Wastes

The wastes from agricultural products are a major pollution hazard. They include run-off or effluent from sawmills, pulp manufacture, fruit and vegetable canning, cleaning of dairies, slaughtering of meat animals, tanning, manufacturing of cornstarch and soya protein, sugar refining, distilling, wool processing etc. Plant diseases move from agriculture to lawns, gardens, parks and golf courses.

POLLUTANTS DAMAGING TO AGRICULTURE

Soil and water pollutants that adversely affect agriculture include sediment, out of place plant nutrients, inorganic salts and minerals, organic wastes, infectious agents, industrial and agricultural chemicals, and heat.

SEDIMENT

Sediment is an out of place resource or contaminant whose dual effect is to deplete the land from which it came and impair the quality of the water it enters. Sedimentation causes blockage of irrigation canals, dams, farm ponds and impairs the oxygen balance in water.

PLANT NUTRIENTS

Plant nutrients become serious pollutants when out of place in ground and surface water. Unwanted aquatic plants are nourished by plant nutrients derived from agricultural run-off, feedlots and barnyards, municipal and rural sewage, and industrial wastes. Aquatic plants clog irrigation and drainage structures and increase nitrates and nitrites in groundwater, which can poison humans and livestock.

PESTICIDE MANAGEMENT POLLUTION AND CONTROL

Prediction of water quality impacts of pesticides and related land management practices is an essential element of site-specific control options and for the development of generic approaches for pesticide control. Prediction tools are mainly in the form of models, many of which are contained in Table. Also, the key hydrological processes that control infiltration and run-off, and erosion and sediment transport, are controlling factors in the movement of pesticides.

EUROPEAN EXPERIENCE

The Netherlands National Institute of Public Health and Environmental Protection (RIVM, 1992) concluded that "groundwater is threatened by pesticides in all European states. This is obvious both from the available monitoring data and calculations concerning pesticide load, soil sensitivity and leaching... It has been calculated that on 65% of all agricultural land the EC standard for the sum of pesticides (0.5; μ g/l) will be exceeded. In approximately 25% of the area this standard will be exceeded by more than 10 times..."

In recognition of pesticide abuse and of environmental and public health impacts the European countries have adopted a variety of measures that include the following (FAO/ECE, 1991):

- Reduction in use of pesticides (by up to 50% in some countries).
- Bans on certain active ingredients.
- Revised pesticide registration criteria.
- Training and licensing of individuals that apply pesticides.
- Reduction of dose and improved scheduling of pesticide application to more effectively meet crop needs and to reduce preventative spraying.
- Testing and approval of spraying apparatus.
- Limitations on aerial spraying.
- Environmental tax on pesticides.
- Promote the use of mechanical and biological alternatives to pesticides.

Elsewhere, as for example Indonesia, reduction in subsidies has reduced the usage of pesticides and has increased the success of integrated pesticide management programmes.

PESTICIDE REGISTRATION

Pesticide control is mainly carried out by a system of national registration which limits the manufacture and/or sale of pesticide products to those that have been approved. In developed countries, registration is a formal process whereby pesticides are examined, in particular, for mammalian toxicity (cancers, teratogenic and mutagenic effects, etc.) and for a range of potential environmental effects based on the measured or estimated environmental behaviour of the product based on its physico-chemical properties. Most developing countries have limited capability to carry out their own tests on pesticides and tend to adopt regulatory criteria from the developed world.

As our knowledge of the effects of pesticides in the environment accumulates, it has become apparent that many of the older pesticides have inadequate registration criteria and are being re-evaluated. As a consequence, the environmental effects of many of the older pesticides are now recognized as so serious that they are banned from production or sale in many countries. A dilemma in many developing countries is that many older pesticides (e.g. DDT) are cheap and effective. Moreover, regulations are often not enforced with the result that many pesticides that are, in fact, banned, are openly sold and used in agricultural practice. The dichotomy between actual pesticide use and official policy on pesticide use is, in many countries, far apart.

Regulatory control in many countries is ineffective without a variety of other measures, such as education, incentives, etc. The extent to which these are effective in developed versus developing countries depends very much on (1) the ability of government to effectively regulate and levy taxes and (2) on the ability or readiness of the farming community to understand and act upon educational programmes. The fundamental dilemma remains one of accommodating local and short term gain by the farmer (and manufacturer and/or importer) by application of an environmentally dangerous pesticides, with societal good by the act of limiting or banning its use.

There is now such concern over environmental and, in some instances, human health effects of excessive use and abuse of pesticides, that there is active discussion within many governments of the need to include a programme of pesticide reduction as part of a larger strategy of sustainable agriculture. In 1992, Denmark, the Netherlands and Sweden were the first of the 24 member states of the OECD to embark upon such a programme. The Netherlands is the world's second largest exporter of agricultural produce after the United States. In contrast, wood preservatives in the forest sector account for 70% of Swedish pesticide use with agriculture using only 30%. As noted above, the lack of baseline data on pesticides in surface waters of

OECD countries, is a constraint in establishing baseline values against which performance of the pesticide reduction programme can be measured.

DANISH EXAMPLE

In 1986 the Danish Government initiated an Action Plan for sustainable agriculture which would prevent the use of pesticides for two purposes (WWF, 1992):

- Safeguard human health - from the risks and adverse effects associated with the use of pesticides, primarily by preventing intake via food and drinking water.
- Protect the environment - both the non-target and beneficial organisms found in the flora and fauna on cultivated land and in aquatic environments.

The objective was to achieve a 50% reduction in the use of agricultural pesticides by 1997 from the average amount of pesticides used during the period 1981-85.

This was to be measured by:

- A decline in total sales (by weight) of the active ingredients
- Decrease in frequency of application. While the World Wide Fund for Nature report that by 1993, sales of active ingredients had been reduced by 30%, the application frequency had not declined.

The Danish legislation included the following components although, by 1993, not all had achieved comparable success.

Pesticide Information: One source of current information on pesticides is through the Extension TOXicology NETwork operated by the University of Oregon on the Internet.

Extoxnet provides information on:

- What is the EXTension TOXicology NETwork?
- Pesticide Information Profiles
- Toxicology Information Briefs
- Toxicology Issues of Concern
- Factsheets
- News about Toxicology Issues
- Newsletters
- Resources for Toxicology Information
- Technical Information
- Mailing Groups
- Search ALL Extoxnet areas for keywords
- Search ALL Extoxnet areas for partial words email access: Almanac@sulaco.oes.orst.edu (request extoxnet catalogue)
- *Reassessment of active ingredients:* Reassessment reflects improved scientific knowledge of pathways, fate and effects of pesticides. By 1993, 80% of the 223 active ingredients had been reassessed. Fewer

than 40% had been approved and about 15% are restricted to specific types of application.

- *Promotion of organic agriculture:* The legislation included funding to promote conversion of traditional agriculture to organic agriculture which, by definition, does not use pesticides.
- *Excise tax on pesticides*: The Danish Institute of Agriculture concluded that, "A tax on pesticides can be designed and implemented in such a way that it will reduce the use of pesticides without distorting or dramatically worsening the economic situation in the agricultural sector." Funds raised by the tax were to be directed back to the agricultural sector. Studies reported by the Institute of Agriculture suggested, however, that pesticide taxes alone would not produce the requisite reduction during the lifetime of the plan.
- *Certification of pesticide users*: All farmers and commercial sprayers must hold application certificates. Certification includes education in pesticide issues.
- *Records of pesticide application*: Commencing 1 August 1993, individual farmers were required to maintain records of pesticide application.
- *Approval of spraying equipment:* This measure gives the Ministry of Agriculture some control of types of spraying equipment used in Denmark. New computer controlled sprayers permit continuous monitoring of pesticide dose by the farmer and reduces excessive application.

The Danish Government is considering the following additional components as part of the regulatory process:

- *Maximum limits on the environmental load of pesticides:* The intent is to produce an index which equates the quantity used of a pesticide with its known ecological effects. The concept is, however, difficult to implement as noted by the WWF, "... there is no direct relationship between the pesticide-load index and the environmental effects - direct or indirect - of pesticides, since these are the result of a complex interaction between many different factors." Nevertheless, the concept has certain management and regulatory value and may be possible, initially, with a few common pesticides.
- *Prohibiting the use of pesticides within 10 m of lakes, watercourses, wetlands, and conservation areas:* This would achieve some level of pesticide protection for aquatic systems in the same manner that buffer strips are widely used to reduce the effects of sedimentation.
- Prohibiting the use of pesticides within a specified distance from private gardens and properties containing fields that are cultivated without the use of pesticides.
- Prohibiting the use of pesticides within 10 m of a drinking-water reservoir.

The Swedes have had considerable success in achieving pesticide reduction targets. WWF credits their success to the following factors.

- Setting of targets with achievable goals and using multiple measures of reduction.
- Lead role played by the Environment Ministry and Chemicals Inspectorate.
- Active support of farmers organizations which realise the economic and environmental advantages of reduced pesticide usage.
- A strong research and development base that provides credible support for new pesticide initiatives.
- Certification of new machinery and routine testing of farm sprayers at government-regulated test centres.
- Re-evaluation and re-registration of pesticides which has resulted in 338 products being removed from the market.

PESTICIDES AND WATER QUALITY

Use of pesticides in developing countries is extremely variable, from nil in large parts of Africa, to extremely heavy dosage in intensive agricultural areas of Brazil and plantations of Central America. In their review of the limited research literature on pesticide use and impacts in Africa, Calamari and Naeve conclude that, "The concentrations found in various aquatic compartments, with few exceptions are lower than in other parts of the world, in particular in developed countries which have a longer history of high pesticide consumption and intense use. Generally, the coastal waters, sediments and biota are less contaminated than inland water environmental compartments, with the exception of a few hot spots."

The Brazilian State of Paraná is typical of developing countries undergoing rapid expansion of agriculture, and illustrates the dilemma of pesticide monitoring. Andreoli reports that Brazil, in general, had become the world's third largest user of"agrotoxic" substances by 1970, only exceeded by France and the United States. However, Andreoli states that less than 15% of active ingredients marketed in Brazil are analysed owing to the lack of methodology, equipment and financial resources. Yet, in a major study of 17 agrotoxic substances (including 11 organochlorine pesticides) over the period 1976 and 1984 in the Paraná River basin, 91.4% of in situ [presumably ambient] samples contained at least one In the Pirapó sub-basin 97.2% of ambient water supply samples and fully 100% of samples from springs showed pesticide residues.

Andreoli also notes that studies of"intoxication" in 1985 showed a predominant effect from organophosphate pesticides, with the most serious effects influencing persons between the ages 15 and 25. The International Code of Conduct on the Distribution and Use of Pesticides, formulated and being implemented by FAO (Box 8) is very relevant to pesticide pollution control and environmental protection in general. The problems of pesticide management in developing countries is somewhat different than those of the developed countries.

These are summarized as:

- Inadequate legislation and enforcement of pesticide regulations, including importation, use and disposal.
- Gifts of pesticides from donors that encourage inefficient use and abandonment of older quantities of the same pesticide.
- Stockpiling of pesticides, especially in countries with unstable governments, leading to abandonment of stockpiles in situations of insurrection and civil war. Examples exist where such a situation led to severe groundwater contamination and public health crises due to dumping of pesticides by untrained civilians.
- Storage and handling is a major problem, including leakage from old barrels and deliberate dumping of surplus pesticide mixtures into water courses following application.
- Destruction of old stores of pesticides (due to deterioration of the active ingredient) is financial prohibitive (estimated at US$ 5000 per tonne) especially as stocks must be moved to a developed country for destruction. Consequently, old barrels deteriorate with leakage into surface and groundwater and/or dumping of stocks.
- Lack of training of users in pesticide handling and application, leading to improper application with environmental and public health consequences.
- Use of pesticides for inappropriate purposes, such as killing of trash fish.
- Use of old pesticide drums for drinking water, cooking, etc.

International Conde of Conduct on the Distributiona and use of Pesticides: This Code of Conduct, adopted by FAO and its member countries in 1985, recognises that:"In the absence of an effective pesticide registration process and of a governmental infrastructure for controlling the availability of pesticides, some countries importing pesticides must heavily rely on the pesticide industry to promote the safe and proper distribution and use of pesticides. In these circumstances foreign manufacturers, exporters and importers, as well as local formulators, distributors, repackers, advisers and users, must accept a share of the responsibility for safety and efficiency in distribution and use."

Prior Informed Consent (PIC) is an important component of the Code of Conduct. Under PIC,"pesticides that are banned or severely restricted for reasons of health or the environment are subject to the Prior Informed Consent procedure. No pesticide in these categories should be exported to an importing country participating in the PIC procedure contrary to that country's decision...". Implementation of PIC is carried out jointly by FAO and the

International Register of Potentially Toxic Chemicals (UNEP/IRPTC) and included 127 countries in December, 1994.

CHANGES OF ATMOSPHERIC PRESSURE

Although atmospheric pressure, or weight of the air per unit horizontal area directly above the place where the pressure is measured, is not a weather element of any particular importance so far as our feelings are concerned, such as temperature and humidity are, nevertheless it is the element that more than any other locates the weather of today and tells us what tomorrow's will be, and therefore is of great importance. Latterly this importance has appreciably decreased, however, because other methods have come into prominence for finding the positions, magnitudes, speeds and directions of travel of the contrasting masses of air whose conflicts and whose passage keep the weather in ceaseless ebb and flow from fair to foul and foul to fair. Still these other methods have not yet been sufficiently developed to justify ignoring the distribution of atmospheric pressure, and how it is increasing here and decreasing yonder, as aids to a foreknowledge of the weather of the morrow. Neither is it at all likely that so great a development and use of these other means, while entirely possible, will be effected at an early date, if indeed they ever are so developed. Therefore a knowledge of the distribution of atmospheric pressure, the near equivalent of the distribution of the air itself, and of its changes, still are, and for a long while must remain, matters both of great practical importance and of keen theoretical interest.

Barometre Corrections

Since the pressure of the air at any particular time and place is, as stated, the weight of all the atmosphere above a horizontal unit area then and there it follows that a knowledge of the values of this pressure over a given region, however extensive, tells us, when the proper corrections have been made, just what the distribution of the air itself is over that same region. But what are these correc- tions? That depends on the kind of instrument used to measure the pressure and on what we want to find, pressure or quantity of air, for we must not forget that the same quantity or mass of air, as also of other things, varies in weight, or pressure, with position not with every change of position but with most changes. But we can think easier and straighter if we consider one thing at a time, and know exactly what that one thing is. Let us first, then, compare the *masses* of air over different places. We can do this by means of a mercurial barometre which is only a device for balancing the pressure exerted by a column of mercury at its bottom, tending to make it flow out, against the pressure there of the atmosphere that keeps it back in. Clearly the pressure exerted by the mercury at the bottom of the column, when corrected for the capillary action of the tube, is proportional to the product of its density by the height of the column by the force of gravity, or pull oil a unit mass by the earth, at that place. Similarly the pressure there of the atmosphere is proportional to the weight, or mass times the force of gravity, of all the air, if still, above that level (the level of the bottom of the mercury column) per horizontal square inch, or other specified area. When the air is in motion, however, as it always is, a further correction of the reading of the barometre is indicated, since the weight of a given mass varies with the direction and speed of its motion. But since the value of this correction (never known accurately) at most is negligibly small it will not be discussed here just mentioned for the sake of completeness.

Similarly the effect of the variation of gravity with height also is negligible. In this case then we may consider sauce for the goose as sauce for the gander too, in the sense that gravity applies in equal (almost) measure and exactly alike for both the mercury and the air. Hence the mass of the air above any given horizontal area, such as a square foot, say, at the level of the mercury in the barometre basin, is proportional to the height of the mercury column, or reading, as we say, of the barometre, corrected for whatever slight difference in that height there may be due to the pull, or capillary action, between the mercury and the wall (glass) of the barometre tube, and corrected also for the density of the mercury corrected to a common or standard density, commonly that which it has at the melting point of ice, 32° Fahrenheit, or 0° Centigrade.

In this way the readings of all the mercurial barometres are made strictly intercomparable, and the ratios of their several readings to each other exactly the same (to within an unknown but always negligible amount) as that of the masses of air above them, so that a geographic map on which these values, if simultaneous, are written, or otherwise represented, is also a map of the distribution of the atmosphere over that same region at the time the readings were made. To obtain the instantaneous distribution of the air the barometre readings necessarily must all be taken at that particular time, whatever it is, 8 A.M. 75th meridian time, or what not.

If, however, aneroid barometres are used instead of mercurial for obtaining the mass distribution of the atmosphere, then their readings, which indicate pressures directly, must be corrected for the local values of gravity, because the atmospheric pressure, being the weight per unit horizontal area of a vertical column of air, varies directly with the force of gravity which, in turn, increases with latitude. This quantity, or mass, distribution of the atmosphere is of considerable importance, especially through its influence on the rate of transmission, in or out, of radiation. A thin atmosphere and a thick one are to the earth much like a sheet and a blanket, respectively, to one sleeping in the open. Although this mass distribution of the atmosphere is very important as a factor that helps to make the weather, especially as to temperature, what it is, still very little use ever is made of it. The distribution of pressure, or push, in the air is on the contrary frequently, carefully and extensively plotted because it is the unequal distribution of push on the atmosphere (or pressure, for they are equal to each other) that makes the winds to blow and thereby, in large measure, but not wholly, change the weather of one place by bringing to it the air and the weather of another. Clearly, then, a knowledge of this distribution must be, and is, a great help to the fore- caster when judging of the weather of the morrow and the day after. That is why the barometre is so extensively used in making meteorological observations and used almost exclusively for measuring the atmospheric pressure. Now the pressure exerted by a column of mercury, the column in this case that is balanced by the pressure of the air, is proportional to the height of the column (corrected for capillary drag), to the density of the mercury (which in turn varies with the temperature) and to the local force of gravity. Hence, to compare the *pressures* at different places, one with another, it is necessary to correct the readings of the barometres to what they would be if all the instruments had the same temperature, say the temperature of melting ice, and were actuated by the same force of gravity its value at sea level at latitude 45°N or some other agreed-upon value.

The readings of aneroid barometres require no corrections for this purpose since, as previously stated, these instruments measure pressure directly. To sum up: If we are measuring the mass distribution of the air the readings of the mercurial barometre must not be altered for gravity differences, while those of the aneroid must be so changed. On the other hand, when it is the distribution of pressure we are finding, the element we nearly always seek, it is the readings of the mercurial barometre that must be corrected to what they would be under some common force of gravity and not those of the aneroid barometre—they are true pressure values, if the instruments are correct, as they stand, and therefore must not be changed. The next and final step is to reduce all these pressure values, or corrected readings of the barometre, to what they would be if taken at some common level. This is done because the travel or flow of the air, due to difference of pressure, is, of course, along the route of least resistance, which route, as a rule, is over a nearly level surface, that is, a surface everywhere normal to the direction of the force of gravity, such as the mean surface of the ocean.

Actually the direction of least resistance to a sample of air through the surrounding atmosphere, the direction indeed of no resistance at all, save alone that incident to viscosity, is along what the physicist calls an isentropic surface. However, this exact and refined treatment of the subject will not here be followed to its conclusion, though to do so affords both surprises and mental satisfaction. But let us have as little boggy ground as possible under our mental feet. Why is the level surface usually the one of least resistance to a sample of moving air? Because ordinarily the change of temperature of the atmosphere with change of height is such that work would have to be done on the sample to carry it to a greater height or push it to a lower level, but no work (save that due to viscosity which is the same in all directions) to move it horizontally.

Normally this distribution of temperature is such that a sample of air taken upward would cool, incident to its expansion under a decreased pressure, to a lower temperature than that of the then adjacent air. Being at the same pressure as this adjacent air but cooler, it also would be denser, so that an actual lift would be required to carry it above its initial level. On the other hand, if taken downwards the compression due to the increased mass of air above it would warm it to a higher temperature than that of the other air at its new position. But being now warmer than the adjacent air, and having the same pressure, it also must be lighter. Hence, normally, only an actual downward push can force a sample of air at any height to a lower plane. In short, air normally can be taken out of its own level only at the cost of work, work against gravity if pushed up, and work against the buoyance of the denser air beneath if forced down. Clearly then no work against gravity, nor any against buoyancy, is required when, and only when, air is moved over a level surface, or, to be exact, over an isentropic surface—a surface which, as already explained, usually is very nearly level. Obviously any direction over this surface is a direction of least resistance.

Therefore if we would know the magnitude of the cause of the winds, and from that magnitude infer their strength, we must find the values of the atmospheric pressure at the *same* level at various places over the region in question. If we had a great number of these pressure readings somewhat evenly distributed over an area of considerable extent, say the size of the United States, or larger, we then could draw on a map of that region a line connecting all the places that, at a particular instant, had the same pressure. This line, according to present practice, we would call all "isobar," or line of equal pressure. By drawing a number of such isobars, corresponding to small but constant differences in pressure, we construct a map or diagram from which the horizontal pressure gradient, or horizontal push per mile, say, can be found. Note that no two isobars (lines of equal pressure) can cross each other. Such crossing would signify two different pressures at the same time and place, a manifest impossibility.

The direction of the flow of the air under these unequal pressures alone would be horizontal, as previously explained, and along the course of the shortest distance between the lines of equal pressure, that is, normal to these lines. A sample of air would not flow off slopingly to this normal to the isobars, though that too would be going from a greater to a lesser pressure, because every component of pressure directed to the right along or parallel to an isobar is exactly balanced by a component of pressure in the opposite direction. The only unbalance is strictly normal to the direction of the isobars, and therefore this is the direction of flow, or would be, if there were nothing to disturb or prevent such flow. As a matter of fact this flow is modified by the rotation of the earth, and so profoundly modified as to change it from the direction of the push to the right angles thereto, except in the lower air where the drag of the surface makes the change in direction less than a right angle. But this fascinating paradox must be passed by. Its explanation would be out of place here; besides only the language of mathematics properly fits it anyway. Since the movement of the air under differences of pressure is horizontal, or very nearly so, as just explained, it is clear that our isobars must be drawn for a common level if we would know from them what the values of the pressure gradients are that are making the winds of the moment to blow, fixing the intensity of a passing storm and largely determining whither and how fast it shall go. But what should be this common level, and how can we find the value of the pressure there? These questions are old; their answers would be new. For the mariner they are simple enough: sea level is the place, and read your barometre where you are (on that level) is the way. But for the landlubber it is a different and awkward matter, since in general no two of his observing stations are at the same level. For the purpose of weather forecasting one can chart on his map the departures of the barometre readings of the several stations from their respective normal values for the season, or day, in question, and then connect by a smoothly curved line the places of equal, like departures. Such a map can be, and has been, used with fair success, but it never became a general favourite, though it is the only pressure map that deals with known accurate magnitudes. One thing that renders such a map less helpful to the forecaster than at first one might expect it to be is the fact that the normal, or average, pressure for the given month, say, is one thing for weather of one kind and another for weather of a different kind.

In short, the general average or normal for all days is not applicable alike to all sorts of weather conditions—the normal itself, or the observed departure from it, needs to be corrected for each special reading, but by how much we do not know. The pressure-departure map therefore is not entirely satisfactory and can not be made so. Another promising procedure, and one that has been seriously attempted, is to use the pressures that exist at some definite height above sea level, half a mile, one mile, or what not. Wherever this level is above the surface, surely at that level there is air. Surely, too, instruments can be carried to that level and the actual pressure there measured with considerable accuracy. However, this direct manner of obtaining the desired pressure values is now too expensive to be practicable.

Therefore the method used has been to compute the pressure at a given height from the pressure and temperature at the surface with a general formula and then correct the value found according to the circumstances, especially as to season and direction of the wind, as empirically deduced from data obtained under various conditions by kite, balloon and airplane. But at best more or less doubt as to accuracy always attaches to the values thus obtained. Hence weather maps that ostensibly show the distribution of atmospheric pressure over a horizontal surface at some definite height above sea level have not yet come into general use. There remains, in this connection, one other obviously possible procedure, the one that has been, and now is, all but universally and solely used. This is, to reduce each pressure reading made on land to what it presumably would be at sea level immediately beneath the point of observation if all below that point, and round about, were not rock and soil, but air in that state and condition appropriate to the location, season and kind and distribution of the weather. The calculation, only approximate of course, is based on the known height of the station above sea level, and the observed temperature and pressure of the air, and then empirically more or less corrected. The chief source of error is the local surface temperature. Sometimes this is abnormally high and sometimes abnormally low. In the first case the reduction to sea level is made on the assumption that the hypothetical air beneath the place of observation is correspondingly warm and light, and in the second that it is correspondingly cold and heavy.

The one yields a pressure too small and the other a pressure too great. Occasionally this error is so big that even the sense of the pressure distribution is reversed—actual cyclones and anticyclones, as indicated by the direction of the winds, changing by reduction to anticyclones and cyclones, respectively, on the map. Pity then the forecaster. But don't pity him overly much. He doesn't deserve it. He just winks the other eye, having in mind any poor devil trying to learn the game from such charts, mentally alters the map in accordance with his background of experience and all data available, some of which are never charted, and goes ahead serenely, if not always confidently. At his own game, and with loaded dice, of course he beats the innocent novice—until he, too, gets wise.

The one thing we definitely know about all reductions of the barometre to sea level, or any other level, is this: They invariably are in error. Generally, however, they are usable, and we do not know of any practical way to improve them. We know too that a "level" surface is not the correct surface on which to draw isobars; but then the surface that is the proper one, the isentropic surface, does not "stay put," as we say, but perpetually so warps and waves as to make pressure reduction to it quite out of the question. As a rule, however, this correct surface is, as previously stated, nearly level. Hence, and because reduction of pressure to a common height usually is only slightly erroneous, it follows that the sea-level map meets the needs of the weather forecaster fairly well, and better indeed than any other that as yet it has been practical to construct. In this connection let us confess a little inconsistency and confusion of practice in our sea-level reductions. We reduce the pressure, or barometre, readings to sea level, but we leave the temperatures unchanged. Hence our weather maps show sea-level pressures and "mile-high" temperatures (themselves often exaggerated by surface conditions) side by side. This for the forecaster. When the climatologist sums the weather data all up he commonly reduces both temperature and pressure to sea level. Consistency, thou mayest be a jewel, but art not always prized.

Index